英語と日本語で学ぶ
熱力学

Ruggero Micheletto　[著]
戸坂 亜希

共立出版

Message from Author

The idea of this book was born in 2011 because of the necessity of a proper "Thermodynamics" text book for the students in the Material Sciences Faculty in Yokohama City University, Japan. This university is highly internationalized and has a number of foreigner teachers that teach sciences courses in English.

Japanese and foreigner exchange students follow classes, but since majority of them are Japanese we matured the idea of the necessity of an appropriate text book in both languages. We contacted the Editor and the reaction was enthusiastic.

The "Thermodynamics" course is given by one of the authors in the autumn semester and lasts about 22 hours (15 classes of 90 minutes). It gives the fundamental of thermodynamics, demonstrates from scratch the ideal gas law, the kinetic theory and Boltzmann principles, teaches what is Brownian motions, its central laws and implications, goes through phenomena like thermal drift and thermal noise, explains key concepts like the thermodynamics fundamental laws, thermodynamical machines and entropy and even touches electrical phenomena and demonstrate the Ohm law.

The class finishes where quantum mechanics begins. At the end of the course the last lectures introduce the limitations of thermodynamics, explain and derive step by step the very first quantum mechanical equation: the Planck "Black Body" radiation formula.

Thermodynamics can be a rather cold and sometimes boring subject. So we focused to keep reader alert and motivated. In this book we never give concepts from above, but incite students to understand the fundamental meaning of things in the real world. We try always to give reasons why those concepts were developed and explain what practical problems a theory solves.

For example, the ideal gas law, usually given as a starting proposition in many textbooks, it is instead derived gradually here from basic concepts of first-year physics. Or, as another example of our approach, after introducing and deriving Boltzmann distribution, we apply it to a column of air and show how it explains everyday life problems like diminished pressure with height or the weaker concentration of oxygen at higher elevations.

At the end of this course students are satisfied from the feeling that they really understood the physical meaning of each formula learned. Instead of forcing students to just learn and memorize abstract things in order to pass an exam, we focused on transmitting the essence of the meaning of things, and give the students the ability of deriving their own physical model so they feel they are really in control of physical concepts.

We hope this book will transmit the same feeling as our course. Yokohama City University score each year in the top 10% as student satisfaction and top 1% in students employment-ratio in Japan, surpassing several Japanese world class universities.

Ruggero Micheletto, PhD in Physics
Yokohama, July 2015

著者からのメッセージ

　皆さんは英語は得意ですか？　英語はあまり得意じゃない，という理科系の学生は少なくありません．ところが，実は，理科系でも研究室に入ると，もれなく英語の論文を読むという輪講やゼミが待ち構えています．ひょっとすると，「ただでさえ英語が苦手なのに，難しい研究の話を英語で読むなんて…」と思うかもしれません．あるいは，「英語は得意だけど，物理の話を英語で読むのは敷居が高そうだ」と思う人もいるかもしれません．
　この本は，そんな皆さんのために書きました．
　いろいろな使い方ができるように，工夫しました．例えば，右のページの日本語のところだけを読むと，熱力学の本になっています．具体的な例を使って，なるべくわかりやすく説明しています．英語が苦手な人は，初めは右のページだけ読んでもよいでしょう．左のページには，右ページとまったく同じ内容が英語で書かれています．英語が苦手だな…と思っている人も，日本語で内容を理解できたのなら，きっと英語でも理解できると思います．
　科学の世界で使われる英語は，実はそんなに難しくありません．なぜなら，科学の世界では論理を明確に伝える必要があるために，比喩や暗喩，難しい熟語などがあまり使われないからです．ごく簡単な文法だけで，科学の英語を読むことはできるのです．
　ただし，普通の英語の教科書には出てこない，いわゆる科学用語やしきたりといったものは覚えなくてはいけません．英語のページには，私がかつて学生だったころに，間違ったことや疑問に思ったことを脚注で解説しました．単に英語のことだけではなく，物理の世界で用いられるしきたりについても書きました．

科学の世界の英語って，ちょっとした専門用語を覚えたらそんなに難しくないんだ！　と皆さんが自信をもつことで，新しい世界がひらけますように．

2015 年 7 月
戸坂亜希

Contents

Chapter 1 Basics for thermodynamics *2*
   1.1 Introduction *2*
   1.2 First experiments with gases *4*
   1.3 The law of ideal gases *10*
   1.4 Isothermal compression/expansion *20*
   1.5 Adiabatic compression/expansion *22*

Chapter 2 A mathematical exercise *30*
   2.1 The physical meaning of differentials *30*
   2.2 How to obtain the circumference of a circle? *34*
   2.3 The area of a sphere *36*

Chapter 3 Laws of thermodynamics *42*
   3.1 What is thermodynamics? *42*
   3.2 First law: the heat engine *42*
   3.3 The second law: order and disorder *44*
   3.4 Entropy *48*

Chapter 4 Distributions *56*
   4.1 What is a "distribution"? *56*
   4.2 The distribution density in an air column *56*
   4.3 The Boltzmann law *66*
   4.4 The concept of distribution of a general potential *74*

Chapter 5 Various phenomena explained by thermodynamics ... 86
5.1 Physical states: gas, liquid and solids ... 86
5.2 Speed distribution in an ideal gas ... 92
5.3 Brownian motion ... 98
5.4 Thermal noise ... 110
5.5 Evaporation ... 116

Chapter 6 Applications ... 130
6.1 Diffusion process ... 130
6.2 The drift velocity ... 138
6.3 Electric resistance ... 142
6.4 Diffusion ... 148
6.5 Black body radiation ... 156

Index ... 169

目 次

第1章 熱力学の基礎 ……………………………………………… *3*
- 1.1 はじめに ……………………………………………………… *3*
- 1.2 気体分子に関する初期の研究 …………………………… *5*
- 1.3 理想気体の状態方程式 …………………………………… *11*
- 1.4 等温圧縮と等温膨張 ……………………………………… *21*
- 1.5 断熱圧縮と断熱膨張 ……………………………………… *23*

第2章 数学のエクササイズ ……………………………………… *31*
- 2.1 微分の物理的な意味とは？ ……………………………… *31*
- 2.2 円周の長さの求め方 ……………………………………… *35*
- 2.3 球の表面積の求め方 ……………………………………… *37*

第3章 熱力学の法則 ……………………………………………… *43*
- 3.1 熱力学とは何だろう？ …………………………………… *43*
- 3.2 熱力学第1法則 …………………………………………… *43*
- 3.3 熱力学第2法則：秩序と無秩序 ………………………… *45*
- 3.4 エントロピー ……………………………………………… *49*

第4章 分　　布 …………………………………………………… *57*
- 4.1 分布とは何だろう？ ……………………………………… *57*
- 4.2 気柱の中の空気の密度分布 ……………………………… *57*
- 4.3 ボルツマンの法則 ………………………………………… *67*
- 4.4 一般的なポテンシャルの分布についての概念 ………… *75*

第5章　熱力学から考える様々な現象 ……………………………… *87*
5.1　気相・液相・固相という物理的な相について ……………… *87*
5.2　理想気体の速度分布 …………………………………………… *93*
5.3　ブラウン運動 …………………………………………………… *99*
5.4　熱雑音 …………………………………………………………… *111*
5.5　蒸発 ……………………………………………………………… *117*

第6章　応用編：熱力学的な概念を使って ………………………… *131*
6.1　拡散のプロセス ………………………………………………… *131*
6.2　ドリフト速度 …………………………………………………… *139*
6.3　電気抵抗 ………………………………………………………… *143*
6.4　拡散 ……………………………………………………………… *149*
6.5　黒体放射 ………………………………………………………… *157*

索　　引 ……………………………………………………………………… *171*

This book is to learn a very important subjects in physics: Thermodynamics. This apparently boring topic is instead taught here in a fun and thought-provoking way.
Remember that Newton though that heat was a sort of liquid and that Brown looking at what we call now Brownian motion thought that inert particles were alive!
In this book you will learn how scientists discovered the fundamentals of all physics and modern sciences: the nature of heat, the existence of atoms, the properties of gas and molecules and up to the quantization of energy.
You will learn these things in a graceful way, step-by-step here.
So please sit back and enjoy studying!

Chapter 1 Basics for thermodynamics

1.1 Introduction

Thermodynamics is a very important branch of physics, it deals with the behavior of many particles and tries to use mathematics in order to control and estimate the global behavior of a great number of independent objects (molecules, particles, charges... you name it!). For example, if we consider the movement of a single molecule in a gas, we can imagine that its movement is chaotic, that the particle will act like a ball colliding randomly with other molecules or other objects around it. Of course we know very well the basic physics of such movements. A straight line, a collision and another straight line again. This is easy Newtonian physics.

But, the problem is that it is clearly too complex to use standard physics to describe an ensemble of a great number of molecules. If we wanted to do so, we would need a number of equations equal to the number of molecules involved. This means, for just a mole of gas, about 10^{23} equations! Clearly this is absurd. Then how can we explain collective molecular phenomena like the temperature, the pressure of a gas, or other collective phenomena like diffusion, Brownian motion or electric conduction? These are the problems that thermodynamical approaches and solves with very good approximations in many cases.

In this book we try to explain the fundamentals of thermodynamics. Especially we want to make the student to understand how some solutions are achieved. We want to give the method to reach a solution in order that he can derive it again by him/herself without the need of the book. Or better, we hope this book will give students a new "way of thinking" and to develop his/her own ability to derive thermodynamical solutions for general

第1章 熱力学の基礎

1.1 はじめに

　熱力学は物理の学問のなかでも，とても大切な分野である．というのも，沢山の粒子の振る舞いについて扱うものであり，数学を用いて沢山の物体（分子，粒子，電荷・・その他何でも）の包括的な振る舞いについて，調べたり見積もったりするためのものだからである．例えば，あなたが気体中の1つの原子の動きについて考える時，その動きは無秩序で，その分子がまわりの気体分子と衝突しながらランダムに動くボールのようなものを想像するのではないだろうか．私たちは，まっすぐに進んで他の分子と衝突し，その後もまっすぐに進むという基礎的な物理についてはよく知っている．これはとても簡単なニュートン物理学である．

　しかし，問題は，沢山の数の分子の記述を考える時には，ごく基礎的な物理法則で記述できる運動でもとても複雑になってしまうということである．その沢山の数の分子の運動を考えようとすると，分子の数だけ方程式を考えなくてはならない．これはつまり，たった1モルの気体だとしても，10^{23} もの式を考えなくてはならないということである！　明らかにこれは無理な話である．それでは，どうやって温度や圧力，拡散，ブラウン運動，電気伝導といった集合した分子の振る舞いを考えればよいのだろうか．これが熱力学の扱うべき問題であり，そしてまた，様々な場合，非常によくこれらの現象を解き明かす術なのである．

　この本のなかで，私たちは熱力学の基礎的な事柄について説明する．なかでも，学生の皆さんが「どうやってその考えを導くのか」といった点に重点を置いた．皆さんが，教科書を見なくても自分の力で答えにたどり着くような力を身につけることを期待している．あるいは，自分の力で「どうやって考えるか」という方法を皆さんに提示し，これを応用して他の一般的な新しい問題について，自分の力で答えを導いて欲しいと思う．

and even new problems.

The book is based on the "Thermodynamics" course (named also "Micro phenomena") in Yokohama City University. We accept both Japanese and International students, so English vocabulary is simplified and we tried to avoid complex phrases in order to be more easily understood by international readers. A full translation with rephrasing and comments about English expressions is also provided on the opposite side of each page for the Japanese student.

1.2 First experiments with gases

Since many years ago people studied the properties of gases. Even in the late 1600, Mr. Boyle discovered that the product Pressure by Volume PV[1] is always a constant, if the amount of gas and its temperature do not change. The experimentalist at the time of Boyle (1650), could measure the expansion with an apparatus that is represented in principle in the figure 1.1:

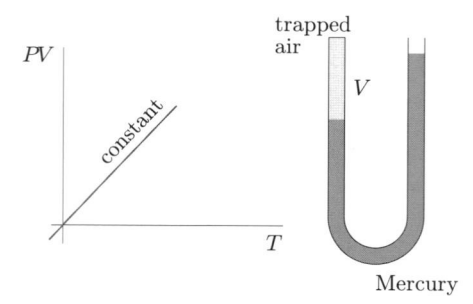

Fig. 1.1 A schematics of the apparatus used by Boyle in 1662. In the graph it is represented the linear behaviour with temperature, however Boyle didn't know that $PV = NkT$ at that time. He only knew that the product PV was a constant, if T was constant, independently by gas composition or nature.

---

1) 基本のルール：物理量はイタリック（斜め文字）で表記する．レポートを書く時には物理量に気をつけて斜体にしてみよう．単位は立体（普通の文字）を使う．

この本は横浜市立大学の熱力学の授業に基づいて書かれている．授業は日本人と，そして外国人の両方のためのものである．英語の語彙は複雑な言い回しではなくごく平易なものを使用し，英語を母国語としない外国人にもわかりやすいように努めた．反対側のページには（つまり，こちらのページ）日本人の学生のために，日本語で書いてある．

1.2　気体分子に関する初期の研究

　遥か昔から，人々は気体の性質についての研究を行ってきた．1600 年代の後半という遠い昔でさえ，ボイルは気体の量や温度が変化しない場合，物質の圧力と体積の積が常に一定であるという法則を発見した．1650 年のボイルの実験は，図 1.1 のような気体の膨張を測定することができる装置を用いて行われた．

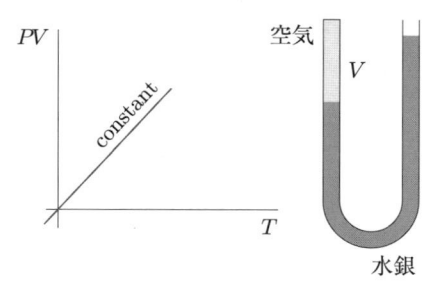

図 1.1　1662 年，ボイルによって用いられた装置の模式図．左のグラフは，圧力と体積の積が温度に対して線形に比例する関係を示している．ボイルはこのとき $PV = NkT$ の関係は気づいていなかった．ボイルはただ気体の組成や性質によらず圧力と体積の積が一定になると気がついていただけだった．

A fixed amount of air was trapped inside this J shaped tube. Boyle changed the amount of mercury in the tube in order to exert more or less pressure on the air. With this simple system he could measure both the air volume and the pressure exerted by the mercury on it. At constant temperature, he was surprised to find the famous relation $PV = constant$. More surprisingly he discovered that this was true for any kind of gas he put inside his J shaped tube! There was a linear relation between the product[2] PV and the temperature T. Boyle noticed that the inclination of the line was approximately the same, for any gas, if he inserted about the same number of molecules. And this was true for ANY gas they tried! Why this was happening? This fact, now called "Boyle law", it is amazing. Why it is amazing? Because of its simplicity! This result is so simple that it is easy to miss what is quite remarkable about it.

Gases come in many different forms. We might have a very light gas like helium, the gas used to lift balloons, whose molecules are little spheres. Or we might have a denser gas like the oxygen of the air, whose molecules are dumbbell shaped. Or we might have a vaporized liquid, like water vapour, whose molecules are shaped something like the letter "Y". In every case, the same law holds, even if the oxygen or water vapour are mixed up with another gas like nitrogen in the air. Even though nothing in the law takes note of all these differences, still this law works...!

Later in the 1700 century J. Charles and J.L. Gay-Lussac discovered that there was some kind of linear dependence with temperature, now we call these laws the Charles law ($V/T = constant$) and the Gay-Lussac law ($P/T = constant$).

---

2) product は積のこと．ちなみに和：sum，差：difference，商：quotient.

この図について詳しく説明しよう．まず，ある一定量の空気がアルファベットのJの形をした管に閉じ込められており，水銀の量を調節することで，空気に働く圧力の大きさをコントロールできるようになっている．この単純な仕組みを使って，ボイルは空気の体積と空気が水銀から受ける圧力を測定した．一定の温度条件での測定結果から，ボイルは有名な $PV = $ 一定 の関係を発見して驚いた．そしてもっと驚くことに，J管に閉じ込める気体の種類を変えても，その法則は成り立ったのであった．

圧力と体積の積 PV と温度 T の間には，線形に比例するという関係がある．ボイルはこの比例係数について，気体の分子量を揃えてJ管に閉じ込めると，どの気体についても同じだということに気がついた．どんな気体を試してもそうなったのだ．なぜだろうか．ボイルが見つけたこの法則は，まさに現在「ボイルの法則」と呼ばれるもので，素晴らしい発見だった．何が素晴らしいかというと，その単純さである．この関係はとてもシンプルなので，私たちは通常この法則が非常に注目に値するものだということを忘れがちである．

気体分子は様々な形態をとる．例えば，風船を浮かすのに使われているヘリウムはとても軽くて，その分子の形は小さな球形のようになっている．空気中に存在する酸素のように集まっている気体もある．酸素分子はダンベルのような形をしている．水蒸気はアルファベットのYの字のような形をしている．ボイルの法則は，分子の形が違っていても同じ法則が成り立ち，また，大気のように様々な形の分子の酸素や窒素が混ざり合っている場合にも成り立つ．つまり，その気体分子の形の違いを知らなくても，この法則は成り立つ．

17世紀後半，J. シャルルとJ.L. ゲイ=リュサックは，温度に対して線形の依存性があることを発見した．これは現在私たちが，シャルルの法則（圧力が一定の場合に $V/T = $ 一定）あるいはゲイ=リュサックの法則（質量と体積が一定の場合に $P/T = $ 一定）と呼ぶものである．

Example 1.1

Suppose you are Boyle in 1662. You trap in the "J" shaped tube a fixed amount of air. Changing the amount of mercury, you change the amount of pressure that is exerted on the gas (air). You first measure pressure and volume and you obtain these values:

$$V_1 = 1 \text{ cm}^3 (= 10^{-6} \text{ m}^3)$$
$$P_1 = 0.1 \text{ MPa} \tag{1.1}$$

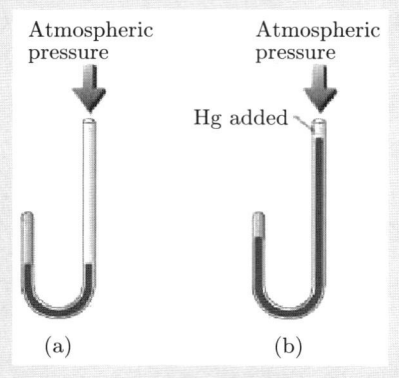

Fig. 1.2 A schematic of the apparatus used by Boyle in 1662. You change the amount of mercury on the column to change the pressure on the gas (panel a and b). Keep PV constant and calculate the new pressure using MKS units

Then you add more mercury in the column. The weight of the column is now the equivalent of 380 mm of mercury over the atmospheric pressure, about 0.05 MPascal, so you have:

$$P_2 = 0.15 \text{ MPa}$$

can you calculate the volume you expect in this case? Please use MKS units[3], Pascal is already in MKS units (1 Pa = 1 Newton per one square meter[4]).

Answer: $V_2 \approx 6.67 \times 10^{-7}$ m^3

例題 1.1

あなたが1662年のボイルだと想像して，考えてみよう．Jの形の管に，ある量の気体を封じ込めた．水銀の量を変えることで，閉じ込めた気体に加わる圧力を変えることができる．まず最初に，圧力と体積を測ると，

$$V_1 = 1 \text{ cm}^3 (= 10^{-6} \text{ m}^3)$$
$$P_1 = 0.1 \text{ MPa} \tag{1.1}$$

という値を得た．

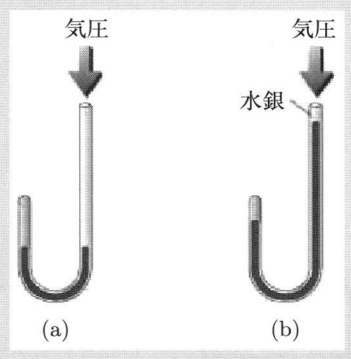

図1.2 ボイルが1662年に用いた実験装置の模式図．気体にかかる圧力を変えて，水銀の量を変えることができる ((a) および (b) 参照)．PV を一定に保ち，変化させた圧力を MKS 単位系を使って考えよう．

次に，先ほどの状態からさらに水銀を加えた．380 mm の水銀柱の管の重さ分の圧力が，閉じ込められた気体にかかっている．この圧力分は，およそ 0.05 MPa なので，閉じ込められた気体の今の圧力は

$$P_2 = 0.15 \text{ MPa}$$

であるとわかる．このとき，体積はいくつになっているだろうか？ ただし MKS 単位系を使うこと（Pa はすでに MKS 単位．$1 \text{ Pa} = 1 \text{ N/m}^2$）

答え：$V_2 \approx 6.67 \times 10^{-7} \text{ m}^3$

1.3 The law of ideal gases

To put together all these relations found by these researchers, we have to understand one of the oldest mystery of science: what is the physical meaning of temperature? We know that material can be hot or cold at touch. But what does that means in physical terms? How do we determine the real nature of the temperature?

Well, we consider a piston full of gas, like in figure 1.3. If we suppose that inside this piston there are N particles, what is the force exerted on the piston? Well, the answer is easy. By definition, if the piston has an area A, the pressure on it is

$$P = \frac{<F>}{A} \qquad (1.2)$$

where the symbolism $<F>$ means *the average force F*.

Now, for sure we can measure the average force on the piston $<F>$, but

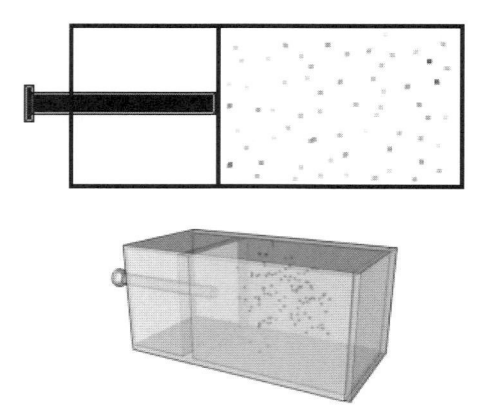

Fig. 1.3 A gas piston with gas molecules in it.

---

3) MKS units: MKS 単位系. メートル (m), キログラム (kg), 秒 (s) を基本単位とする単位系のこと. 科学論文では MKS 単位系を使うのが一般的.
4) square: 2乗のこと. ここでは m² という意味.

1.3 理想気体の状態方程式

ボイル，シャルル，ゲイ=リュサックによって発見された法則をまとめるにあたって，まず私たちは科学の最古の謎について理解する必要がある．それは，物理学的な見地に立った時に温度の意味とは何かというものだ．私たちは物質に触れば，当然それが熱いか冷たいかを知ることができる．けれども，物理的にはどういう意味があるのだろうか．温度の本質とは何なのかをどうやって結論づけたらよいのだろうか．

まず，図1.3のようにシリンダーに気体分子が一杯につまった場合について考えよう．ピストンの中にN個の気体分子が入っていたとすると，ピストンに働く力はどうなるだろう．答えはとても簡単だ．ピストンの面積がAだとすると，ピストンに作用する圧力は

$$P = \frac{<F>}{A} \tag{1.2}$$

となる．ここで$<F>$は平均の力のことである．

ピストンに働く平均の力$<F>$の測定は簡単に可能である．しかし，個々の分子がピストンに衝突する時の力はどうだろうか．まず，1個の分子の運動量について考えてみよう．運動量は

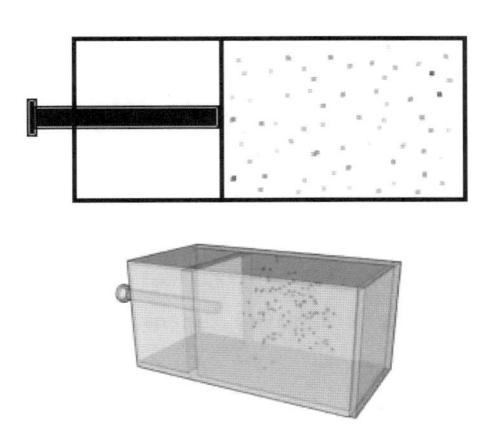

図1.3 気体分子が入っているピストンの模式図

how much it is the *single* force of a single molecule hitting the piston? Let's consider the *momentum* of a particle, it is

$$p = mv_x \qquad (1.3)$$

we indicate v_x instead of v because we now consider only the component of the force orthogonal to the piston surface. We call this orthogonal direction x.

What is the net force acting on the piston? Naturally, it is the force of a single particle multiplied by the total number of particles hitting the piston, lets call this number N^*, so: $<F> = N^*F$. Let's now estimate this force. We remember from basic physics that the impulse $p = mv_x$ is related to the force by

$$F = \frac{p}{t} = \frac{mv_x}{t} \qquad (1.4)$$

where t is an interval of time. This is the force of a single hit, a collision of a molecule on the piston. Let's find how many hit will occur in a small amount of time t. Well, only the particles *near enough* to the piston will hit it, these particles are those at a distance less than $v_x t$. All the others they do not have enough time to reach it (!) These particles are enclosed in a volume $V^* = Av_x t$, where A is the area of the piston wall. Now, if we keep proportion in our mind, of course the ratio of the number of particle hitting the piston against the total particles, and the ratio between the volume that enclose them against the total volume, should be the same... so this relation must be true:

$$\frac{N^*}{N} = \frac{V^*}{V} = \frac{Av_x t}{V} \qquad (1.5)$$

so the total number of particles hitting the piston in the time t is:

$$N^* = N\frac{Av_x t}{V} \qquad (1.6)$$

Now using the fact that $<F> = N^*F$ and equation 1.4, we can calculate:

1.3 理想気体の状態方程式

$$p = mv_x \tag{1.3}$$

である．ここで v ではなく v_x と置いたのは，ピストンに垂直に衝突する分子のみを考えるためである（表面に垂直な方向を x とした）．

ピストンに働く正味の力はどうだろうか．自然に考えると，ピストンに衝突する個々の分子の力を足し合わせたものになるだろう．ピストンに衝突する分子の数を N^* とすると，その力は $<F>=N^*F$ となる．ここで，この力について，見積もってみよう．基本的な物理学を思い出してみると，力積 $p=mv_x$ は力と関係があり，

$$F = \frac{p}{t} = \frac{mv_x}{t} \tag{1.4}$$

という式で表すことができる．ここで t は時間である．この力は 1 個の分子がピストンに衝突したときの力である．では短い時間 t の間に何回衝突があったかについて考えよう．時間 t にピストンに衝突する分子は，ピストンからの距離が $v_x t$ よりも小さいところにあると考えられる．なぜなら，$v_x t$ よりもピストンから遠くにいる気体分子は，ピストンに衝突することができないからである．つまり，衝突する分子がいる場所の体積は $V^* = A v_x t$ となる（A はピストンの面積）．ここで，頭の片隅に，ピストンに衝突する分子の数と全分子数の割合は，ピストン近くの体積と全体の体積の割合と，同じであることを覚えておいてほしい．つまり，

$$\frac{N^*}{N} = \frac{V^*}{V} = \frac{Av_x t}{V} \tag{1.5}$$

となり，時間 t の間にピストンに衝突する全分子数は

$$N^* = N \frac{Av_x t}{V} \tag{1.6}$$

となる．

$<F>=N^*F$ と式 (1.4) より，

$$<F> = N \frac{Av_x t}{V} \frac{mv_x}{t} \tag{1.7}$$

と計算が可能で，面積 A で割ると，

$$<F> = N\frac{Av_x t}{V}\frac{mv_x}{t} \tag{1.7}$$

now we simplify t and divide by the area A, we obtain immediately that:

$$PV = Nmv_x^2 \tag{1.8}$$

Now we notice that we have a term $mv_x^2 \ldots$, this is twice the kinetic energy $1/2 mv^2$. Let's try to substitute this kinetic energy in our equation (1.8). To do so, we must be careful. Until now we considered the sole x direction, however for general random and uniform velocities, the kinetic energy is distributed equally in all directions, so it is

$$<v_x^2> = <v_y^2> = <v_z^2> = \frac{1}{3}<v^2> \tag{1.9}$$

this means that the total kinetic energy is three times[5] $\frac{1}{2}$. So finally we can write:

$$PV = \frac{2}{3}N<E_k> \tag{1.10}$$

where $<E_k> = mv^2/2$.

Again we should remember that the brackets $<>$ represent the *average* value. Coming back to our experiments, we know that the product PV is proportional to the temperature multiplied[6] by some constant. Very interestingly if we look well at equation (1.10), and compare it with $PV = constant \times T$, it is clear: the temperature seems to be something proportional to the kinetic energy of our gas! This lead to a profound result, we solved one historical question: *what is the nature of temperature?* Temperature is a physiological sensation due to the kinetic energy of the molecules that collide and hit our body. More energy makes us feel hot, less energy feel cold.

Now we need only to clean up a little equation (1.10). We define

---

5) three times：3 倍.
6) multiply O_1 by O_2: O_1 に O_2 を掛ける.

1.3 理想気体の状態方程式

$$PV = Nmv_x^2 \tag{1.8}$$

を得ることができる.

ここで, mv_x^2 という項があることに気がついただろうか. これは運動エネルギー $1/2mv^2$ の 2 倍である. では, 運動エネルギーを式 (1.8) に代入してみよう. ただし, このとき, 私たちは慎重にならなくてはいけない. というのもこれまで私たちは一方向, x 方向だけを考えてきたからである. 一般的に, ランダムで等しい速度をもつ場合には, 運動エネルギーはどの方向でも同じ値をもつべきである. つまり,

$$<v_x^2> = <v_y^2> = <v_z^2> = \frac{1}{3}<v^2> \tag{1.9}$$

となる. これは全運動エネルギーは各方向の 3 倍で, それぞれは全方向の $\frac{1}{3}$ だということである. このことから, 私たちは最終的に

$$PV = \frac{2}{3}N<E_k> \tag{1.10}$$

と書くことができる. ここで $<E_k> = mv^2/2$ である.

繰り返しになるが, ここで $<>$ という記号は平均ということを表している. 私たちの (頭の中での) 実験について思い出してみると, 私たちは圧力と体積の積 PV が温度に対してある定数で比例するということを知っていた. また, 大変面白いことに, もう一度式 (1.10) を注意深く見て, そして $PV = constant \times T$ と比べると, 温度というのが気体の運動エネルギーに比例する関係であるということがわかる. これは大変興味深い結果である. つまり, 私たちは一つの歴史的な謎「**温度の本質とは何だろうか**」という問題を解いたということだからである. 温度というのは, 私たちの体に衝突する分子の運動エネルギーを生理学的に感じるものなのである. エネルギーが高いほど私たちは熱いと感じ, エネルギーが低いほど冷たく感じるのである.

式 (1.10) を整理してみよう. そうすると, 以下の式

$$<E_k> = \frac{3}{2}kT \tag{1.11}$$

を定義することができる. ここで k はボルツマン定数である. そうすると,

$$<E_k> = \frac{3}{2}kT \qquad (1.11)$$

where we call k the Boltzmann constant, we have the nice and famous expression:

$$PV = NkT \qquad (1.12)$$

This equation is called usually *the ideal gas law*. Experimentally k results to be equal to $k = 1.38 \times 10^{-23}$ Joule per degree Kelvin. (If we want to use another constant, instead of the number of molecules N, we can use the number of *moles* n. Then the eq. (1.12) becomes $PV = nRT$ with R the *universal gas* constant $R = 8.31$ J/mol K.) Remember that we chose eq (1.11) just because it is convenient not to have the number 3 or 2 in the final formula eq. (1.12). However, this was just a choice that was made years ago, any other definition would have been good as well.

Example 1.2

Suppose we have a rubber balloon full of Air. We are at room temperature $T = 300$ K, its volume is one litre and its pressure is a little more than the atmospheric pressure, let's say 1.5 atmospheres. Can we estimate the total number of molecules N that are inside the balloon?

Yes it is possible, we simply use the equation $PV = NkT$, and extract N from that. In these basic calculations we must be careful not to make mistakes with dimensions. Let's do it together putting all dimensions between square brackets [ ][7].

We use MKS. First we consider that 1 litre is 10^{-3}m^3 and that 1 atmosphere is about 0.10 MPa. Now we write

---

7) 計算するときに単位を考えるというのはとても大事なポイント．熱力学だけではなく，他の計算でも，計算するときには単位を書こう．

有名でよく使う式,

$$PV = NkT \tag{1.12}$$

を導くことができる．この式は通常，**理想気体の状態方程式**と呼ばれている．経験的に k は 1.38×10^{-23} J/K である．もしも他の定数を使いたい場合，たとえば N の代わりに分子数をモルで表記したい場合には n を使う．そうすると式 (1.12) は気体定数の $R = 8.31$ J/molK を用いて $PV = nRT$ となる．私たちは式 (1.11) を最終的な式 (1.12) にするとき，3 や 2 といった数字を入れたくなかったために選んだ．しかし，他の定義を用いてもいけないわけではない．

例題 1.2

空気で膨らませた風船を考えてみよう．その風船は，室温 $T = 300$ K，体積 1 リットル，圧力は大気圧よりも少し高くて，1.5 気圧くらいだとしよう．その風船の中にいる分子数 N を見積もってみよう．

これはとても簡単で，$PV = NkT$ を使えば見積もることができる．ただし，単位次元について間違えやすいので，すべての単位を [ ] の中に入れて一緒に計算しよう．

MKS 単位系を使うので，まず 1 リットルを 10^{-3} m^3，1 気圧を 0.10 MPa と直そう．すると

$$N = \frac{PV}{kT} = \frac{0.15 \times 10^6 [\text{Pa}] \times 10^{-3} [\text{m}^3]}{1.38 \times 10^{-23} [\text{J}]/[\text{K}] \times 300 [\text{K}]} \tag{1.13}$$

となる．ここで，単位について慎重になろう．ジュール [J] という単位は MKS 単位系では $[\text{J}] = [\text{m}^2 \times \text{kg/s}^2]$ なので，

$$N = \frac{PV}{kT} = \frac{0.15 \times 10^6 [\text{Pa}] \times 10^{-3}[\text{m}^3]}{1.38 \times 10^{-23}[\text{J}]/[\text{K}] \times 300[\text{K}]} \quad (1.13)$$

we have to be careful with units: considering that the unity [J] is in MKS $[\text{J}] = [\text{m}^2 \times \text{kg/s}^2]$ and that $[\text{Pa}] = [\text{kg}]/([\text{s}^2][\text{m}])$, we can rewrite:

$$N = \frac{PV}{kT} = \frac{0.15 \times 10^6 \frac{[\text{kg}]}{[\text{s}^2][\text{m}]} \times 10^{-3}[\text{m}^3]}{1.38 \times 10^{-23}[\text{m}^2] * \frac{[\text{kg}]}{[\text{s}^2]*[\text{K}]} \times 300[\text{K}]} \quad (1.14)$$

you note that all dimensions disappear, so at the end we have

$$N = \frac{PV}{kT} = \frac{0.15 \times 10^6 \times 10^{-3}}{1.38 \times 10^{-23} \times 300} \approx 3.6 \times 10^{22} \quad (1.15)$$

This is the number of molecules in the balloon! Can we estimate also the weight of this thing? Yes, why not? If we know how many molecules are there, we can estimate also the total weight. If we assume that Air is roughly 80% of Nitrogen, from the chemical tables we know that one mole of Nitrogen (that is about 6×10^{23} molecules) weights 14 grams. So a mole of air roughly speaking should be about 0.014 [kg]/0.8, that is 0.017 [kg] or 17 grams. Our total number of molecules is not a mole, but much less $N = 3.6 \times 10^{22}$. How much less? The fraction of moles of air we have in the balloon N_{mol} is

$$N_{mol} = \frac{3.6 \times 10^{22}}{6 \times 10^{23}} = 6 \times 10^{-2}$$

so the total weight of the balloon is 17 grams multiplied by N_{mol}, that is

$$weight = 17[\text{g}] \times 6 \times 10^{-2} \approx 1\text{g}$$

As you noticed we did very rough assumptions during this calculation. However, you may agree that - very roughly speaking - this value can approximate the weight of our balloon, at least in the **order of mag-**

$$N = \frac{PV}{kT} = \frac{0.15 \times 10^6 \, \frac{[\text{kg}]}{[\text{s}^2][\text{m}]} \times 10^{-3} [\text{m}^3]}{1.38 \times 10^{-23} [\text{m}^2] * \frac{[\text{kg}]}{[\text{s}^2]*[\text{K}]} \times 300 [\text{K}]} \quad (1.14)$$

とかける．ここで，すべての単位が打ち消し合ってなくなったのがわかるだろうか．つまり，

$$N = \frac{PV}{kT} = \frac{0.15 \times 10^6 \times 10^{-3}}{1.38 \times 10^{-23} \times 300} \approx 3.6 \times 10^{22} \quad (1.15)$$

となる．

これが風船の中にいる原子の個数である．その質量を見積もることができるだろうか．もちろん簡単なことである．風船のなかに何個原子があるかわかれば，すぐに合計の質量を計算することができる．仮に，風船の中の空気の 80% が窒素だとすると，1 モル（およそ 6×10^{23} 個の分子）の質量は 14 g である．つまり 1 モルの空気は 0.014 [kg]/0.8 となり，0.017 [kg] となる．風船のなかの気体は 1 モルではなく，もっと少ない $N = 3.6 \times 10^{22}$ 個である．どのくらい少ないかというとその比は，N_{mol} は

$$N_{mol} = \frac{3.6 \times 10^{22}}{6 \times 10^{23}} = 6 \times 10^{-2}$$

となる．つまり，質量は，

$$weight = 17 \, [\text{g}] \times 6 \times 10^{-2} \approx 1 \, \text{g}$$

となる．この概算はとても粗い計算だ．しかし，非常に荒っぽく言ってしまえば，少なくともその質量は桁では合っているはずだ．あなたはもっときちんとした見積もりを練習として，やってみればよい．

nitude[8]. You can devise a more refined estimation of the weight by yourself as an exercise.

1.4 Isothermal compression/expansion

We will see later that equation (1.12) is also called the equation of isothermal compression, because in the mathematical space PV (where PV are the variables, as XY are variables in the Cartesian space) if temperature is constant we have the relation $PV = const.$ (k is obviously a constant and N is the number of molecules in the box, so it is a constant too). We can plot this curve in the PV space (figure 1.4). This is called an *isothermal process*[9] because it represent the relation between pressure and volume of a gas when its temperature is not changing.

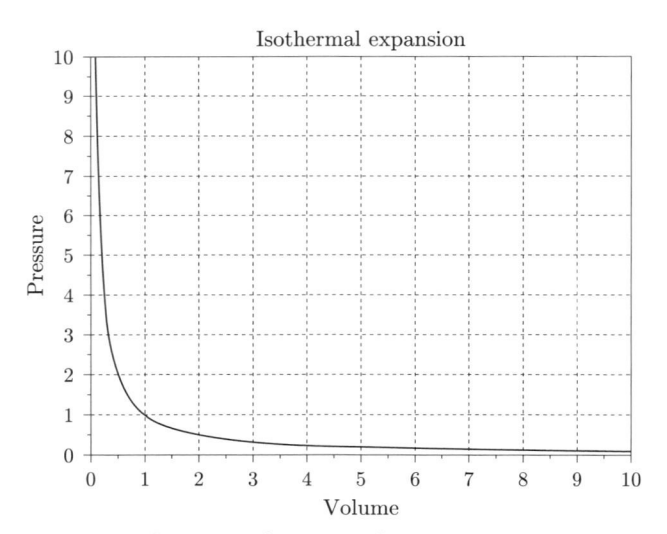

Fig. 1.4 The isothermal process of a gas in the PV space. Please notice that this behaviour is valid for any time of gas or mixture of gases!

---
8) order of magnitude というのは桁という意味. 厳密な値は違うけど桁では等しいなどといった時によく使われる言葉なので覚えておこう.

9) isothermal compression: 等温過程.

1.4 等温圧縮と等温膨張

前の章に出てきた式 (1.12) の $PV = NkT$ は等温圧縮の式とも呼ばれている．なぜなら，数学的空間の PV（XY がデカルト空間での変数であるように PV が変数という意味）は，温度が一定ならば，$PV = $ 一定 という関係がある（k は明らかに定数であり，N は箱に入っている分子の数なので，これも一定である）．つまり私たちは PV 空間においてこの曲線をプロットすることが可能である．そのプロットを図 1.4 に示す．これは**等温過程**と呼ばれている．なぜなら温度が変化しない場合において，圧力と体積についての関係を表しているものだからである．

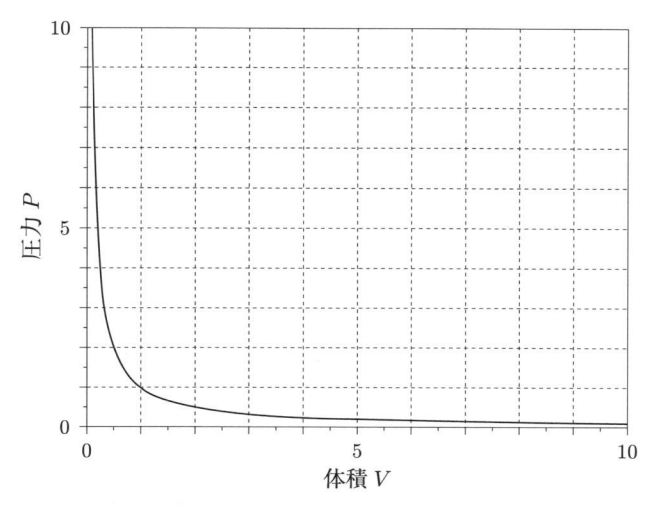

図 1.4　PV 空間での気体の等温過程．この振る舞いはどんな時にも気体の種類によらずに当てはまるということを覚えておこう．

1.5 Adiabatic compression/expansion

Now let's make a final effort and make some more considerations: what happens if we strongly compress a gas in a piston? Of course, the temperature of the gas will increase. So these kind of compressions cannot be *isothermal* as the one above. How will be the dependence of Pressure against Volume in this case?

As we know energy and temperature are related accordingly to equation (1.11). Using (1.10) and (1.11) we can write

$$PV = N\frac{2}{3} < \frac{1}{2}mv^2 > \qquad (1.16)$$

let's consider now the total kinetic energy of the system $U = N < \frac{1}{2}mv^2 >$, this U is generally called the total *internal* energy of a gas. It is the multiplication of all the molecules present, N, by their average kinetic energy.

$$PV = \frac{2}{3}U \qquad (1.17)$$

This equation is valid if all the energy is only kinetic energy[10]. In other words if there are not other forms of *internal* energies, like for example molecular rotation, molecular bending, molecular vibration or other form of molecular energy. If our gas is for example a mono-atomic gas, then all these other forms of energy are negligible and the internal energy U is a good approximation of the total energy within the gas. For historical reasons, let's modify equation (1.17) as

$$PV = (\gamma - 1)U \qquad (1.18)$$

this of course can be always done. It just means that the parameter γ is equal to 5/3 in the case of an ideal mono-atomic gas.

---

10) kinetic energy: 運動エネルギー.

1.5 断熱圧縮と断熱膨張

さて，もう少し考えてみよう．もしも，私たちがピストンの中にある気体を圧縮したら何が起こるだろうか．もちろん気体の温度は上昇する．つまり，このような圧縮の場合，1つ前に勉強した等温変化ではなくなる．では，こういう場合には圧力や体積の変化はどうなるだろうか．

前に勉強したように，エネルギーと温度の関係は式 (1.11) によって明らかになっている．そこで，式 (1.10) と式 (1.11) から，

$$PV = N\frac{2}{3}<\frac{1}{2}mv^2> \tag{1.16}$$

という式を導くことができる．

ではここで，系の全運動エネルギー $U = N<\frac{1}{2}mv^2>$ を考えてみよう．ここで U は一般的に気体の全内部エネルギーと呼ばれている．これはすべての分子についての足し合わせであり，分子の平均の運動エネルギーは以下のようになる．

$$PV = \frac{2}{3}U \tag{1.17}$$

この式はすべてのエネルギーが運動エネルギーだけの場合に有効である．他の言葉で置き換えると，分子の回転や分子の変角振動，分子振動やそのほか分子のエネルギーの様態などといった，内部エネルギーに他の項がない場合ということである．単分子気体の場合について考えると，他の項を無視することができるため，内部エネルギー U が気体の全エネルギーであるという近似はうまくいく．先人の知恵を拝借して，式 (1.17) を

$$PV = (\gamma - 1)U \tag{1.18}$$

のように変形してみよう．これはいつもやる手で，なんでこう変形するかというと，理想的な単原子気体の場合に γ は 5/3 と等しくなるからである．

次に，系にする仕事をどうやって式で表すかということについて考えてみよう．基礎的な物理学から，仕事は Fdx だと知っているので，この積をピストンの面積によって割ったり，掛けたりしてみると，次式のようになる．

1.5 Adiabatic compression/expansion

Now let's suppose we want to calculate an expression for the work of the system. We know from basic physics that Fdx is the work, multiplying and dividing[11] by the area of the piston (figure 1.3), it is straightforward that

$$\frac{F}{A} \times A dx = work \qquad (1.19)$$

The pressure is F/A and Adx is the small variation of volume dV due to the expansion of the piston. This is the mechanical work exerted to the gas by the compression. Where is this mechanical energy (mechanical work is energy!) going to end up? If there are no losses outside, this work will correspond to an equal and opposite variation of internal energy of the gas! If so, we can write $PdV = -dU$. We said above that $U = PV/(\gamma - 1)$, so we have only to differentiate equation (1.18):

$$dU = (PdV + VdP)/(\gamma - 1)$$

Now we substitute $PdV = -dU$ and we have:

$$(\gamma - 1)dU = (PdV + VdP)$$
$$-(\gamma - 1)PdV = (PdV + VdP)$$

Now eliminating PdV we have:

$$\gamma PdV = -VdP$$
$$\gamma \frac{dV}{V} + \frac{dP}{P} = 0$$

Now we integrate and we have:

$$\gamma \int \frac{dV}{V} + \int \frac{dP}{P} = \int 0$$

remembering that the integral[12] of zero is a constant and that $\int dx/x$ is $ln(x/x_0)$ we have:

---

11) divide: 割る.
12) integral: 積分.

1.5 断熱圧縮と断熱膨張

$$\frac{F}{A} \times Adx = work \tag{1.19}$$

F/A というのは圧力で，Adx は dV と同じで膨張によって変化した体積である．つまり，この仕事というのは，圧縮によって気体にする力学的な仕事であると言える．この力学的なエネルギー（力学的な仕事というのはエネルギーということ）は結局どうなるのだろう．もしも外側に失われることがないのならば，この仕事は内部エネルギーの変化の逆と等しくなる．そうすると，$PdV = -dU$ と表すことができる．前に $U = PV/(\gamma - 1)$ と記述をしたが，私たちはこの式 (1.18) を微分すると このようになる．

$$dU = (PdV + VdP)/(\gamma - 1)$$

ここで，$PdV = -dU$ の関係を代入すると，

$$(\gamma - 1)dU = (PdV + VdP)$$
$$-(\gamma - 1)PdV = (PdV + VdP)$$

ということがわかり，PdV を消去するため，

$$\gamma PdV = -VdP$$
$$\gamma \frac{dV}{V} + \frac{dP}{P} = 0$$

という式を積分すると，

$$\gamma \int \frac{dV}{V} + \int \frac{dP}{P} = \int 0$$

となる．ゼロを積分すると定数となり，$\int dx/x$ は $ln(x/x_0)$ となることを思い出すと，

$$\gamma lnV + lnP = const. \tag{1.20}$$

となる．もしもあなたが対数の性質を覚えているならば，この式はつまり，下記のようになる．

$$PV^\gamma = const. \tag{1.21}$$

$$\gamma \ln V + \ln P = const. \qquad (1.20)$$

if you remember the properties of logarithms, this means that:

$$PV^\gamma = const. \qquad (1.21)$$

This is a very important relation because it expresses the behavior of pressure and volume in the case we have an expansion or compression that actually changes the temperature of the gas and there are not external losses of heat. Equation (1.21) it is also called the equation for an *adiabatic expansion* (or compression)[13]. Please notice that the exponent γ is beautifully at the exponent of the volume. Now you understand why in equation 1.18 we used this parameter instead of the usual 2/3.

Actually, we can *measure* γ experimentally, just making an adiabatic expansion and measuring P and V. We find that for mono-atomic gases γ is about 5/3 as expected. Instead for more complex gases, because of vibrational and other spurious phenomena that make the gas a non-ideal one, we can measure different values of γ. The isothermal and adiabatic expansions look different on a PV plot as shown schematically in figure 1.5.

---

13) adiabatic expansion: 断熱膨張, adiabatic compression: 断熱圧縮.

この関係はとても重要な関係である．なぜなら，圧縮や膨張が起こる時の圧力と体積の振る舞いを表すものだからである．先ほど学んだように，まさに気体の温度が変化し，内部での熱の損失はない．式 (1.21) は**断熱膨張**（あるいは収縮）の式と呼ばれる．ここで，指数 γ が体積 V の指数となっていることに注目してほしい．そうすると，私たちは，なぜ式 (1.18) において通常の 2/3 という数値ではなく，このパラメータを用いたのかということが理解できるだろう．事実，私たちはこの γ を，断熱膨張のときに P と V を測定することにより，実験的に明らかにすることができるのである．

実際に実験を行うと，単原子気体の場合では γ が期待通り 5/3 となることがわかる．もっと複雑な気体の場合では，分子の振動や不純物の影響で，気体の振る舞いが理想気体的ではなくなるので，γ の値は変わってくる．等温あるいは断熱の膨張の場合，PV のプロットがどうなるかを図 1.5 に示す．

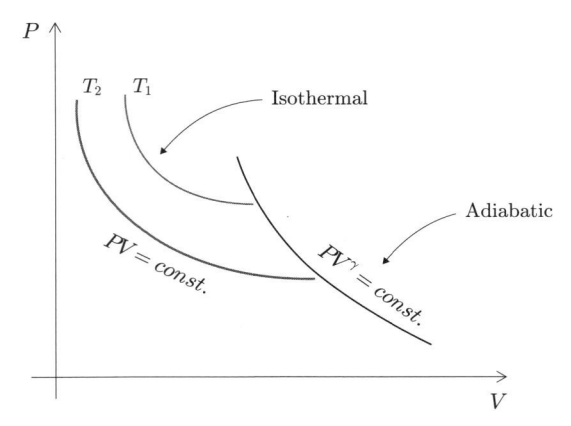

Fig. 1.5 A scheme of isothermal and adiabatic expansions of a gas in the PV space.

Example 1.3

Let's suppose that a gas is kept in a box at constant temperature and at a pressure of $P_1 = 1\text{Atm}$, volume $V_1 = 1$ litre. We let expand the gas to 1.2 litre without temperature control, what happens to the pressure?

Using what we understood above, the product PV^γ must be constant. So we simply calculate the pressure with this expression

$$P_2 = \frac{P_1 V_1^\gamma}{V_2^\gamma}$$

Let's convert this to real numbers, we again pay attention to units, we must use MKS always, so we remember that 1 Atm is about 10^5 N/m^2:

$$P_2 = \frac{10^5 \text{ [N/m}^2\text{]} \times 0.001^{5/3} \text{ [m}^3\text{]}}{0.0012^{5/3} \text{ [m}^3\text{]}} = 7.3 \times 10^4 \text{ [N/m}^2\text{]}$$

This show how a small difference in volume can result on a big drop of pressure in the adiabatic expansion.

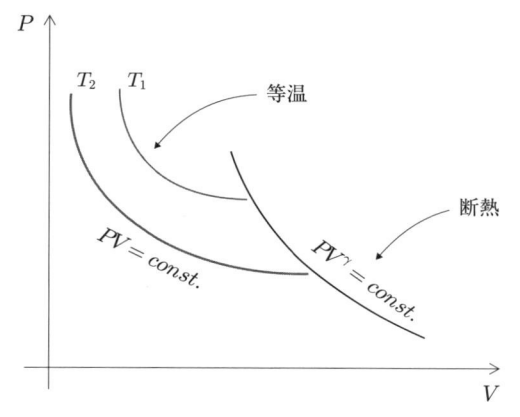

図 1.5　PV 空間における気体の等温または断熱膨張のときのふるまい

> **例題　1.3**
>
> ある体積 $V_1 = 1$ リットルの箱のなかに一定の温度の単原子気体が圧力 $P_1 = 1$ 気圧で閉じ込められていると考えよう．温度を制御せずにこの箱の体積を 1.2 リットルまで広げたとき，圧力はどうなるだろうか．
>
> 上記の通り，PV^γ の積は一定に保たれる．つまり，
> $$P_2 = \frac{P_1 V_1^\gamma}{V_2^\gamma}$$
> という式から簡単に圧力を計算することができる．
>
> 実際の値を入れて考えてみよう．その時は，単位の換算にしっかりと注意しなくてはならない．私たちは普段 MKS 単位系を使うので，1 気圧は $10^5 \, [\text{N/m}^2]$ と覚えよう．すると
> $$P_2 = \frac{10^5 \, [\text{N/m}^2] \times 0.001^{5/3} \, [\text{m}^3]}{0.0012^{5/3} \, [\text{m}^3]} = 7.3 \times 10^4 \, [\text{N/m}^2]$$
> となる．この計算から，断熱膨張ではわずかな体積の違いによって，大きく圧力が異なることがわかるだろう．

Chapter 2 A mathematical exercise

2.1 The physical meaning of differentials

In this course it is very important to understand differential[14] models of the processes we study. For this reason it is of paramount importance to grasp the philosophy of differentiation and integration of a variable. For this purpose let's play around with few basic problems. Trying to solve these, will help us to understand the principles of differentiations from the physicist point of view.

In the real world, physical variables are related in complex way. Differentiate means to consider very small differences of variables in order reduce *curves* in *straight lines*. In other words, we *think small* and reduce complexity. Any curved path, can be considered as the sum of many small linear segments. For very small differences, any complex relation can be considered like a linear one. The key point is that these difference should be really small.

For example let's consider the relation $PV = NkT$ as we studied in the previous chapter. We have three variables (P, V and T) that are related by a defined mathematical relation. If V changes, the other two variables must change accordingly along a certain *curve*. Let's study the small differences between two points of this equation, point one will be (P_1, V_1, T_1) and point two (P_2, V_2, T_2). For example:

$$P_1 V_1 = NkT_1 \qquad (2.1)$$

$$P_2 V_2 = NkT_2 \qquad (2.2)$$

---

14) differential: 微分.

第 2 章 数学のエクササイズ

2.1 微分の物理的な意味とは？

微分のモデルの意味というものを理解することは大切なことである．なぜなら，微分や積分の「哲学」をしっかりと掌握するというのは，何よりも大切なことだからである．そのために，いくつかの基礎的な問題から取り組もう．この問題を解くことにより，微分の本質というものを物理学的な見地から理解できるだろう．

実際の世界では物理の変数というのは複雑に関係しあっている．微分は微小変化する**曲線的な変数**を，**直線的なものさし**で考えようということである．別な言葉で言い換えると，つまりそれは**複雑さを減らす**ということである．どの曲線も短い線形の部分をつなぎ合わせたものだと見なすことができる．また，変化が小さい場合，変数は直線的だと見なすことができる．重要なポイントは，その変化が本当に小さいべきだということである．

例えば，第 1 章で勉強した $PV = NkT$ という関係について考えてみよう．3 つの変数 P, V および T は数学的な関係式で結びついている．もしも V が変化した場合，他の 2 つの変数はある曲線的な関係で変化する．では，ここで，差が少しだけの場合の 2 つの状況を考えよう．1 つ目は P_1, V_1, T_1 という場合，2 つ目は P_2, V_2, T_2 という場合である．

式は以下のようになる．

$$P_1 V_1 = NkT_1 \tag{2.1}$$

$$P_2 V_2 = NkT_2 \tag{2.2}$$

その差をとってみよう．すると，

$$P_2 V_2 - P_1 V_1 = Nk(T_2 - T_1) \tag{2.3}$$

となる．

let's take the difference:

$$P_2V_2 - P_1V_1 = Nk(T_2 - T_1) \tag{2.3}$$

now, let's add $+P_2V_1 - P_2V_1$, this term is zero so we can do that:

$$P_2V_2 + P_2V_1 - P_2V_1 - P_1V_1 = Nk(T_2 - T_1) \tag{2.4}$$

now, we combine P_2 and V_1:

$$P_2(V_2 - V_1) + V_1(P_2 - P_1) = Nk(T_2 - T_1) \tag{2.5}$$

$$P_2\Delta V + V_1\Delta P = Nk\Delta T \tag{2.6}$$

Untill now we didn't do anything special, we just used basic algebra and called the difference of two points with the symbol Δ. Now think about the fact that these two points are very near each other: if the differences between the point 1 and point 2 are really, really small, then this means that the value of P_1 and that of P_2 are almost the same. They are so similar that they can be confused. So why to write P_1 or V_2 or V_1, V_2 if they are practically the same? If they are, let's call them just $P \approx P_1 \approx P_2$ and $V \approx V_1 \approx V_2$ and write again our equation (2.6):

$$PdV + VdP = NkdT \tag{2.7}$$

You see? We obtained, with only few algebra[15] and some simple thinking, that the differential equation of $PV = NkT$ is $PdV + VdP = NkdT$. The symbol d in dV or dP is sometime called *infinitesimal*. We knew that from mathematics. But now we gave to the differentiation process a *physical meaning*. We learned that to remember how to differentiate complex physical equation, you only need to take two points of the equation and *think small*!

Any differential equation, though, is valid only for points so near each

---

15) algebra: 代数 (学).

2.1 微分の物理的な意味とは？

ここで，$+P_2V_1 - P_2V_1$ を足し合わせると，この項はゼロになるため，

$$P_2V_2 + P_2V_1 - P_2V_1 - P_1V_1 = Nk(T_2 - T_1) \tag{2.4}$$

となる．P_2 と V_1 をまとめると，次のようになる．

$$P_2(V_2 - V_1) + V_1(P_2 - P_1) = Nk(T_2 - T_1) \tag{2.5}$$

$$P_2 \Delta V + V_1 \Delta P = Nk \Delta T \tag{2.6}$$

ここまで，私たちは何も特別なことをしていない．普通の代数学と，Δ という記号を使って 2 点の差をとっただけである．では，ここでこの 2 点が非常に近いという点に着目してみよう．もしも点 1 と点 2 の間が，本当に本当にちょっとだとしたら，P_1 と P_2 はほとんど同じだということができる．これらは似ているので，混乱の原因になる．実質的に同じだとしたら，P_2 を P_1 と記述してもよいし，V_1 を V_2 と書いてもよいだろう．そこで，先ほどの式は $P \approx P_1 \approx P_2$ と $V \approx V_1 \approx V_2$ とも表せる．すると式 (2.6) を書き直すと下記のようになる．

$$PdV + VdP = NkdT \tag{2.7}$$

わかっただろうか．私たちはただ代数学を使い，そしてちょっと考えただけで，$PV = NkT$ の微分の式 $PdV + VdP = NkdT$ を導くことができた．dV あるいは dP などに記号 d は「無限小」と呼ばれる．私たちはその意味を数学から知ることができる．しかし，考えたいのは**物理的な意味**である．私たちは，どうやって複雑な物理の式を微分したらよいのか，ということを，単に式の 2 点をとって，そしてその差が小さいと考えただけで導いた．

どの式の微分も，ほとんど一致すると見なせるような近い 2 点について考えているだけで，理論的には意味があるけれども，**積分しないと実用的な応用**には使えない．では実用的になるように，積分についても考えてみよう．皆さんはもちろん，積分というのが何かは知っているだろう．しかし，もう一度言うけれども，実用的な物理の問題に対応するためにも，積分の物理的な意味についていくつかの問題を解いて考えてみよう．

other, that almost coincide... so they do not have so much use in practical applications. To be useful they must be *integrated*.

Clearly you know how to integrate. But again, to understand the physical meaning of integration in a practical physical problem, let's do some more exercises.

2.2 How to obtain the circumference of a circle?

Let's consider a simple application problem: what is the circumference of a circle of radius[16] R? Of course we all know it is $l = 2\pi R$, but let's pretend we do not know that. Let's *think small* and concentrate on a very small angle, like in the figure 2.1. The angle should be so small, that the curvature of the vertical arc can be neglected. We consider the arc as a straight segment of length dl.

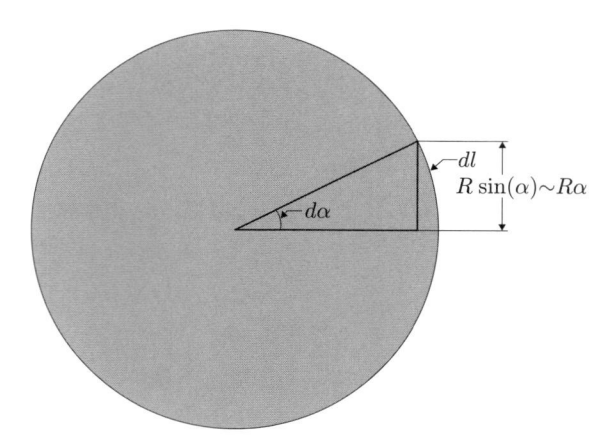

Fig. 2.1 If the angle $d\alpha$ it is an *infinitesimal* then the correspondent arc can be so small that it can be neglected. We considered is as a mere straight segment of a triangle (right).

---

16) radius: 半径, diameter: 直径.

2.2 円周の長さの求め方

シンプルで応用的な問題を考えてみよう．半径 R の円の円周はいくらだろうか．もちろん，$l = 2\pi R$ という関係は知っている．どうしてそうなるかを考えよう．図 2.1 に示すようなとても小さい角度について考えよう．その角度はとても小さいので，円弧の曲率については無視できる．すると円弧は直線 dl であると考えることができる．

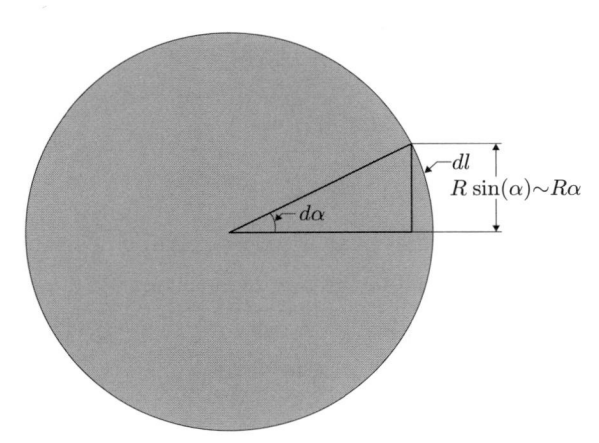

図 2.1 もしも角度 $d\alpha$ が小さいとしたら，その円弧の曲率は無視できる．つまり，右に示すように三角形の一辺として考えることが可能である．

さて，三角形については，三角関数から

We have then a triangle, we know from trigonometry that

$$R \sin d\alpha = dl$$

We call the angle $d\alpha$ because it is as well so small, almost zero, as dl. At this point we have also the advantage that for small angles it is valid the relation $\sin d\alpha \approx d\alpha$[17]. So if we *thought small* enough, we can write with very good approximation

$$dl = R d\alpha$$

But this is not the circumference of our circle, it is just the very small *infinitesimal* part of it!

We have to *integrate* on all the angles and we will have our circumference.

$$\int dl = \int_0^{2\pi} R d\alpha$$

From basic mathematics we know that R can go out of the integral because it does not depend on α, we finally have our circumference:

$$l = \int_0^{2\pi} R d\alpha \tag{2.8}$$

$$l = R \int_0^{2\pi} d\alpha \tag{2.9}$$

$$l = 2\pi R \tag{2.10}$$

This is exactly what we expected! How we do this? We simply found the correct relation for the variables in case of very small differences and then integrated. This is a principle that is very often used in making physical models in thermodynamics and in other fields.

2.3 The area of a sphere

Just to have fun in the same way we can calculate now the area of a

---

17) $\sin d\alpha \approx d\alpha$ というのはよく使う近似だが,この α が何度くらいまで近似が使えるかどうか,具体的な角度について一度考えてみることをお勧めする.

$$R\sin d\alpha = dl$$

という関係を知っている．私たちはこの角度を $d\alpha$ と呼ぶ．とても小さく，dl のようにほとんどゼロだからである．小さな角度の場合には $\sin d\alpha \approx d\alpha$ という関係で変化するという利点も持っている．このような場合，

$$dl = Rd\alpha$$

という関係を導くことができる．しかし，これは円周ではなく，小さな角度をもっている円周の一部としてである．

円周を計算するためには，私たちはすべての角度について積分をしなくてはいけない．積分をすると，次のようになる．

$$\int dl = \int_0^{2\pi} Rd\alpha$$

基礎的な数学知識から，R を積分の外に出すことができる．なぜなら，R は α に依存していないからである．そうすると私たちは円周を得ることができる．

$$l = \int_0^{2\pi} Rd\alpha \tag{2.8}$$

$$l = R\int_0^{2\pi} d\alpha \tag{2.9}$$

$$l = 2\pi R \tag{2.10}$$

これはまさに私たちが求めたかった式である．どうやって導いたかというと，単純にとても小さな差の場合について，変数の正しい関係を見いだし，積分しただけである．これは統計力学や，その他の場合についてよく使われる手法である．

2.3　球の表面積の求め方

同じ手法を使って，球の表面積についても計算してみよう．この場合，計算は二次元になるので，もう1つの垂直方向の角度 ϕ を使おう．前までの結果から，円周は $l = 2\pi R$ で求められることが明らかになった．そこで，今度は

2.3 The area of a sphere

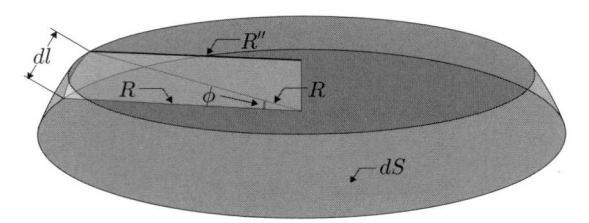

Fig. 2.2 dS is an *infinitesimal* surface formed by the circumference multiplied by the small hight $dl \approx Rd\phi$.

Sphere. This time we are in two dimensions, we must use another angle ϕ that runs vertically. We start from our previous result, the length of the circumference is $l = 2\pi R$. We *think small* again, and we construct a very small surface on our circumference in the vertical direction. To do so, we just multiply the circumference by the small length $dl = Rd\phi$, exactly in the same fashion as before (see figure 2.2).

$$dS = (2\pi R)dl \tag{2.11}$$

But what is the circumference? Is it really $(2\pi R)$ as written above? Well, it is not! If we look at figure 2.2, we notice that while we go up with angle ϕ, the radius R'' gets smaller and smaller. From trigonometry again, we know that $R'' = R\cos\phi$, so the correct relation for the circumference is $2\pi R\cos\phi$, then we have

$$dS = (2\pi R\cos\phi)dl \tag{2.12}$$

$$dS = (2\pi R\cos\phi)(Rd\phi) \tag{2.13}$$

$$dS = 2\pi R^2 \cos\phi d\phi \tag{2.14}$$

the we integrate and we have:

2.3 球の表面積の求め方

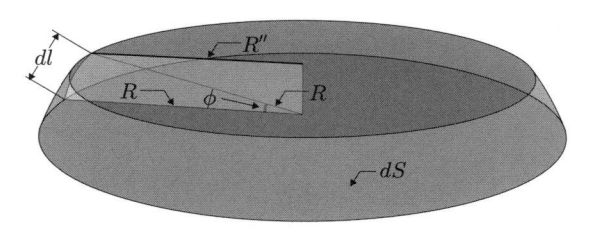

図 2.2 dS は円周と微小高さ $dl \approx Rd\phi$ をかけた無限小の表面積

垂直方向の微小角度を考えよう．そのために，私たちは微小の円周 $dl = Rd\phi$ の足し合わせを考えなくてはいけない．これも前と同じように考えよう（図 2.2 参照）．

$$dS = (2\pi R)dl \tag{2.11}$$

ここで一度，円周とは何かと考えてみよう．さっき書いたように本当に $2\pi R$ なのだろうか．いや違う．図 2.2 を見てみると，私たちは ϕ が大きくなるにつれて，R'' がどんどん小さくなるということに気がつくだろう．三角関数を使って再び考えると，$R'' = R\cos\phi$ となり，円周を $2\pi R\cos\phi$ として修正すると，

$$dS = (2\pi R\cos\phi)dl \tag{2.12}$$

$$dS = (2\pi R\cos\phi)(Rd\phi) \tag{2.13}$$

$$dS = 2\pi R^2 \cos\phi d\phi \tag{2.14}$$

となり，積分をすると

$$\int_0^S dS = \int_0^{\pi/2} (2\pi R)(R\cos\phi d\phi) \tag{2.15}$$

$$S = 2\pi R^2 \int_0^{\pi/2} \cos\phi d\phi \tag{2.16}$$

$$S = 2\pi R^2 [\sin(\pi/2) - 0] \tag{2.17}$$

$$S = 2\pi R^2 \tag{2.18}$$

$$\int_0^S dS = \int_0^{\pi/2} (2\pi R)(R\cos\phi d\phi) \tag{2.15}$$

$$S = 2\pi R^2 \int_0^{\pi/2} \cos\phi d\phi \tag{2.16}$$

$$S = 2\pi R^2 [\sin(\pi/2) - 0] \tag{2.17}$$

$$S = 2\pi R^2 \tag{2.18}$$

If we take a good look again at the figure, integrating from 0 to $\pi/2$ we covered only the top hemisphere, so we have to multiply by two, and at the end we have what they thought us at elementary schools:

$$S = 4\pi R^2$$

Now, the volume of a sphere is $V = \frac{4}{3}\pi R^3$ can you calculate it yourself?

となる．

　ここで再びよく図を見返すと，0 から $\pi/2$ まで積分しているので，私たちが半球しか考えていないことがわかる．つまり結果を 2 倍することで，最終的に小学校で習う

$$S = 4\pi R^2$$

を得ることができる．さて，もう皆さんは，球の体積 $V = \frac{4}{3}\pi R^3$ を自分で導くことができるだろうか？

Chapter 3 Laws of thermodynamics

3.1 What is thermodynamics?

What is thermodynamics? In few words we can say that is the science of studying the relation of between mechanical energy and thermal energy. Thermodynamics is an older theory, and it goes back to the time of Sadi Carnot the beginning of 1800 century, when the molecular theory of matter was not know well. So what we studied up to here, was not yet known. We are going back in time a little now.

3.2 First law: the heat engine

The theory of thermodynamics being so old is based on simple observations, more than rigorous molecular models. At the beginning of the 1800th, the heat was considered to be a kind of fluid that flows in materials, and even the principle of conservation of energy was not established. Let's consider as an example a rubber band. If we expand it, we can observe an increase in the temperature of the band. If we put it between the lips, we can clearly feel that it gets warmer. On the contrary, if we relax it quickly, we can feel that the rubber gets cooler.

These observations made the people of these years to make the following considerations about heat and mechanical force. When a rubber band is holding a weight it is well extended, it is stiff and if we measure the temperature it is hot. On the contrary, when we relax the band, for sure is not holding any weight, it is then not stiff, and we observe it is cool.

People also observed that heating up a rubber band that was holding a weight, actually lifted the weight a little more. Whereas, cooling it down caused the extension of the rubber band, and released the down a weight

第3章 熱力学の法則

3.1 熱力学とは何だろう？

さて，熱力学とは何だろうか．この言葉について，ごく短い言葉で述べるとすると，それは力学的なエネルギーと，熱エネルギーをつなぐものである．熱力学というのは古くからある理論で，それはサディ・カルノーの時代，1800年頃からあった．このとき，分子の理論というものは，よくわかっていなかった．これまで私たちが学んだことはまだわかっていなかったのである．そこで，ちょっと昔にもどって，熱力学についての勉強をしてみよう．

3.2 熱力学第1法則

熱力学の理論は古くから行なわれてきた研究で，簡単な観察に基づいたものである．昔は厳密な分子のモデルはわかっていなかった．1800年代の初め頃，熱というのは流体で物質の中を流れるものだと考えられていた．そして，エネルギーの保存という考え方は，確立されていなかった．ゴムのバンドについて考えてみよう．ゴムのバンドを伸ばすと，バンドの温度は上昇して，唇に挟んでみると，温かくなっていることを感じるはずである．反対に，それを素早く弛めると，ゴムバンドの温度は下がる．

この体験から，人々は熱と力学的な力について，以下のような考えを抱いたのである．それは，ゴムバンドをぴんと張ったとき，ゴムバンドには付加がかかり，その付加が強くなると温度があがる．一方，弛めると付加がかからないので冷たく感じる，というものである．

人々はまた，ゴムバンドを温めると少し縮むことを発見した．一方，冷やすと少しだけ伸びる．そこで得た最初の結論は，物質に熱を加えると，力学的仕事が生じるということである．力学的な仕事というのは，もちろん Fdx を意味する．つまりゴムバンドを力 F によって dx だけ動かした，ということである．

attached to it. So the first conclusion is: if we put heat into a material we can have mechanical work for it. In our example for *mechanical work* of course we mean the product Fdx that we obtain if an object attached to our rubber band is moved up of dx by the force F of the rubber.

All these observations led to general considerations: if a system has a certain *internal* energy U, then any variation of this internal energy must be equal to any external heat ΔQ plus any mechanical work we give to this system. In mathematical terms:

$$\Delta U = \Delta Q + \Delta W \tag{3.1}$$

this is called *the first law of thermodynamics*. In the case the internal energy does not change, we can of course write:

$$\Delta Q = \Delta W$$
$$Q_2 - Q_1 = \Delta W \tag{3.2}$$

But what is the physical nature of this *internal energy U*, and what is the nature of the *heat Q*?. In 1800, it was not clear!

3.3 The second law: order and disorder

In the times of Sadi Carnot, who was the son of a famous military leader, the steam engine was popular and under production, however there was not a proper theory to explain its working principle. So you can understand the importance of the principle of thermodynamics at that time. Carnot was very interested in these machines, and he was able to state a very important principle, the so called second principle of thermodynamics. This principle can be expressed by this phrase:

The heat flows spontaneously from a hot body to a cold one. The inverse process is not possible unless we introduce external work.

これらの実験結果から，一般的な結論を導くことができる．系に内部エネルギー U があったとすると，この内部エネルギーの変化は，系が吸収した熱量 ΔQ とこの系にした力学的な仕事 ΔW の和になる，ということである．

$$\Delta U = \Delta Q + \Delta W \tag{3.1}$$

これは，**熱力学の第1法則**と呼ばれている．内部エネルギーが変化しない系の場合，

$$\Delta Q = \Delta W$$
$$Q_2 - Q_1 = \Delta W \tag{3.2}$$

となる．この内部エネルギー U とはどんな性質をもち，熱 Q とはどんな性質をもつのだろうか．1800年代には，これらはまだ明らかにされていなかった．

3.3 熱力学第2法則：秩序と無秩序

有名な軍人の息子として生まれたサディ・カルノーの時代，蒸気エンジンは一般的に製造されていたが，それが働く原理についてはきちんとした理論は確立されていなかった．そのことを考えると，あなたは熱力学の原理を学ぶことの重要性を理解できるのではないだろうか．カルノーはこの機械について，とても興味をもち，非常に重要な原理を発見した．それは，熱力学の第2法則として知られる法則で，**熱は熱いほうから冷たいほうに流れ，逆は，外部の仕事がないと起こらない**という原理である．

これと同じことを違う方法で表現することもできる．これは，「無秩序」の考え方として知られている．言い換えると「外部からの仕事なしに，**全体としての無秩序を増やすことはできない**」ということである．これら2つの言

There is another way to express the same thing, is with the introduction of the concept of *disorder*. The other way to express the second principle of thermodynamics is then:

In a system any process can only increase the total disorder, unless we introduce external work.

These two phrases express the same things we can observe from real experience. The fact that a hot body put aside a colder one transfers its energy to the cold one and not the opposite.

Why we introduce the concept of *disorder*? Well, we know what is heat. It is something proportional to the molecular speed (remember eq. 1.11). A cold body is a body where the molecules have less speed, so are characterized by a degree of *order* which is lower of a hot body. In a hot body the molecules are moving more chaotically, so there is more disorder.

To clarify this in your mind just think to a cube of ice, a drop of water and a cloud of water vapour. These systems are just the same material at different temperatures, which of these systems contains more order? For sure the cube of ice has more order than the other two. Because the molecules of it are well organized in a cubic crystal. Instead water and vapour molecules move chaotically around each other in all directions. If we want to compare the level of order of these two who wins? Of course water! Why? Because at least water molecules are bind at a certain distance between each other. They move around, but they do not go too far away of each other like in the gaseous water vapour. So the level of order is inversely proportional to the level of temperature. More temperature, less order. So if we leave a cube of ice alone he can melt or sublimate, if we want a droplet of water to become a cube of ice, we have to add external work. These observational statements are the key of the second law of thermodynamics.

The second law of thermodynamics is somehow related not solely to ther-

葉は，同じことを表現しており，例えば，熱い物体を冷たい物体と接触させると，熱は熱いほうから冷たいほうへと移動し，反対は起こらないという実体験を通して観察することができる．

さて，なぜ私は「無秩序」についての話をしたのだろう．私たちは，熱が何であるかということを知っている．熱は分子の速度に比例している（式 (1.11) を思い出そう）．冷たい物体では，分子の速度は遅い．つまり，「秩序」の度合いは，熱い物体のほうが低くなる．熱い物体のなかでは，分子の動きはより混沌としている．つまり，「無秩序」な状態である．

このことについて考えるために，氷のキューブ，水の雫，そして水蒸気について考えてみよう．これらの系では，同じ物質だが温度だけ違う．それぞれの秩序はどうなるだろうか．もちろん，氷のキューブの場合，ほかの2つの状態よりも秩序がある．なぜなら，結晶として分子が並んでいるからである．一方で，水や蒸気の場合，分子は混沌としていて，すべての方向に動き回っている．水と蒸気では秩序の度合いはどちらが高いだろうか．もちろん，水である．なぜなら，少なくとも水の分子は，一定の距離で結合をしているからである．水分子は動き回るが，気体分子に比べて，すごく離れた場所までは動かない．つまり，秩序の程度というのは，温度によって変わる．温度が高くなると，秩序の度合いは低くなる．たとえば，氷のキューブをおいておくと，溶けたり昇華する，あるいは水滴が氷になるということは，外部の仕事を必要とするということでもある．こういう観察というものは，熱力学の第2法則のカギとなるものである．

熱力学の第2法則というのは，熱のプロセスに個別に関係しているもので

mal processes, it has a more general significance. For example let's think of another process, a drop of ink in a glass of water. What happens with time? At first the droplet of ink is very well visible in the middle of the glass. It has a precise position within the glass, it has a certain radius, a certain shape and a certain color. So to describe it we require a great amount of *information*. If we need information to describe a systems, it means we have a certain level of order in it.

Please remember that a completely disordered system, contains very low information. Or in other words, we need very few information to describe it. Just think about white noise. What info you need to describe it, if not that is white noise at a certain volume? Instead think about another sound, the sound of an instrument. How much information you need to describe it...? More than before! You need to say which tone (frequency), what kind of instrument (its timber, its spectrum) and this itself is a very complex set of information. So more information, more order, less information less order.

Coming back to our example of the ink in the glass, after time passes the droplet of ink gets bigger and bigger and at the end it dissolves in the water. What we have after some time, is a glass of water with no apparent droplet of ink, just a glass of water. The color of the water has changed a bit, it is somewhat darker than before. How much information we need to describe this system compared to before? More or less information? Less information! Now to describe our system we have just to say the color of the water...that's it! The information in the system has spontaneously reduced, this means that the total disorder has increased. So this again is in agreement with the second principle, as stated above.

3.4 Entropy

We know now that it is impossible -in a macroscopic system- to reverse the thermodynamical processes, the disorder will always increase. However, what will happen if we *think small*? If we imagine to have a very small sys-

はなく，もっと一般的な意義がある．例として違うプロセスのことを考えてみよう．コップの水にインクをたらした場合に何が起こるだろうか．最初，コップのなかに，インクの固まりがあるのが見えるはずである．一定の大きさ，形，そして色で．つまり，そこには情報がある．ある系についての情報を把握しているということは，秩序の度合いが高いということでもある．

　逆に言うと，まったく秩序がない系は，非常に情報が少ないということを覚えておいて欲しい．白色雑音（ホワイトノイズ）について，考えてみよう．あなたは白色雑音について，どんな情報があるだろうか．白色雑音というのは，周波数に関係なくある一定の大きさであるものである．雑音以外の音について考える代わりに，その装置の雑音についてまず考えなくてはいけない．でも，どれくらい白色雑音について知る必要があるのだろうか．あなたはそのトーン（周波数）だとか，どういう種類の装置だとか（その材質やスペクトルなど）を知らなくてはいけないし，それ自体がとても複雑な情報の組み合わせなのである．つまり，情報が増えると，より秩序が増え，情報がなくなると秩序がなくなるのである．

　グラスのなかのインクの話に戻ろう．時間がたつとインクの粒はどんどん広がり，グラスのなかでインクがどの場所にあるかわからなくなり，ただコップの中の水にしか見えない．水の色は，ほとんど変化せず，ほんの少しだけ色が濃くなる程度だろう．では，前と比べて，あなたはこのコップの中のインクについて，どのくらいの情報を持っているだろうか．情報は前より増えるだろうか，あるいは減るだろうか．もちろん，減るだろう．わかっていることと言えば若干の色の変化ぐらいではないだろうか．情報は自然に減ってしまった．つまり，全体的にみて無秩序が増えたということである．繰り返すと，これは前に述べた第2法則の原則と一致するのである．

3.4　エントロピー

　私たちは，巨視的な系においては，熱力学的なプロセスが，可逆でないことを知っている．無秩序は常に増え続ける．しかし，小さい系で考えたらどうだろうか．とても小さい系で少しだけの分子があり，温度が少しだけ（δT だ

tem with few molecules, and we increase the temperature just a little (δT), is it possible that energy flows also from cold to hot? Well, if we remember that the temperature represents the average velocity of the molecules, if the physical system is small, with very few molecules... maybe! When the system is very very small, and the two bodies have almost the same temperature, of course we can imagine that the temperature always fluctuates from the cold body to the hotter and vice versa freely.

So, from the point of view of *differentials* the second law of thermodynamics is not valid and we can speak of *reversible engines*. (A reversible heat transfer is equivalent to frictionless motion in mechanics.)

Please remember PV plot in figure 1.5 in the first chapter. Let's consider two points a and b at the same temperature. The work done of course is

$$\Delta W = \int_a^b P dV \tag{3.3}$$

Since we are along an *isothermal*, we know that

$$P = \frac{NkT_1}{V} \tag{3.4}$$

so using eq (3.2)

$$\Delta Q_1 = \int_a^b NkT_1 \frac{dV}{V} = NkT_1 ln \frac{V_b}{V_a} \tag{3.5}$$

If we move along another isothermal of course we can write an equivalent equation

$$\Delta Q_2 = \int_c^d NkT_2 \frac{dV}{V} = NkT_2 ln \frac{V_d}{V_c} \tag{3.6}$$

Now if we connect the points b and c, d and a with *adiabatic* curves in the PV plot (see Fig. 3.1), we know from what we already studied that $P_1 V_1^\gamma = P_2 V_2^\gamma$.

Using $PV = NkT$ we have:

3.4 エントロピー

け) 上昇するような場合, エネルギーは, 冷たい系から熱い系に流れたりするだろうか. 温度は分子の平均速度を表しているということを, 覚えているだろうか. ごく小さい系なら, 分子もごく少しだと考えられる. もしも, 系がとても小さくて, 2体がほぼ同じ温度だとしたら, 2体の温度は常に, 冷たくなったり, 熱くなったりと揺らぐだろう.

「差」から考えてみるということは, 熱力学の第2法則が有効ではなくなるので, 可逆機関について話ができるだろう. 可逆であるということは, つまり, 機械に摩擦がないということである.

第1章に書いた, 図1.5のPVプロットを思い出して欲しい. そして, aとbの2点が同じ温度だと考えてみよう. すると, 仕事は

$$\Delta W = \int_a^b P dV \tag{3.3}$$

となる. 等温的な変化なので,

$$P = \frac{NkT_1}{V} \tag{3.4}$$

となり, したがって, 式(3.2)より

$$\Delta Q_1 = \int_a^b NkT_1 \frac{dV}{V} = NkT_1 ln\frac{V_b}{V_a} \tag{3.5}$$

となる. また, 同様に他の等温的な変化について,

$$\Delta Q_2 = \int_c^d NkT_2 \frac{dV}{V} = NkT_2 ln\frac{V_d}{V_c} \tag{3.6}$$

という式を導くことができる.

さてここで, bとc, dとaはPVプロットのなかで, 断熱的な曲線でつなぐことができるが (図3.1参照), 私たちはすでに$P_1V_1^\gamma = P_2V_2^\gamma$を勉強した. $PV = NkT$を用いると,

$$P_1V_1^\gamma = P_2V_2^\gamma \tag{3.7}$$

$$P_1V_1V_1^{\gamma-1} = P_2V_2V_2^{\gamma-1} \tag{3.8}$$

$$NkT_1V_1^{\gamma-1} = NkT_2V_2^{\gamma-1} \tag{3.9}$$

という式が得られる. これは

3.4 Entropy

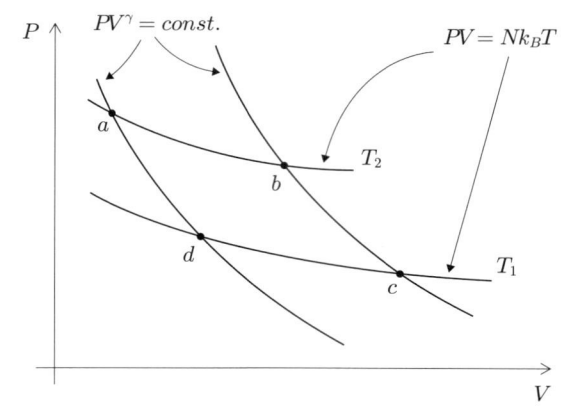

Fig. 3.1 A representation of the isothermal and adiabatic curves mentioned in the text.

$$P_1 V_1^\gamma = P_2 V_2^\gamma \tag{3.7}$$

$$P_1 V_1 V_1^{\gamma-1} = P_2 V_2 V_2^{\gamma-1} \tag{3.8}$$

$$NkT_1 V_1^{\gamma-1} = NkT_2 V_2^{\gamma-1} \tag{3.9}$$

this is equivalent to

$$TV^{\gamma-1} = constant \tag{3.10}$$

This lead to this two equations:

$$T_2 V_b^{\gamma-1} = T_1 V_c^{\gamma-1}$$

$$T_2 V_a^{\gamma-1} = T_1 V_d^{\gamma-1} \tag{3.11}$$

just dividing these two equations we understand that

$$\frac{V_b}{V_a} = \frac{V_c}{V_d} \tag{3.12}$$

This means that the logarithms in eq (3.5) and eq (3.6) are identical, so dividing the two by T_1 and T_2 respectively, we obtain this very important final equation:

3.4 エントロピー

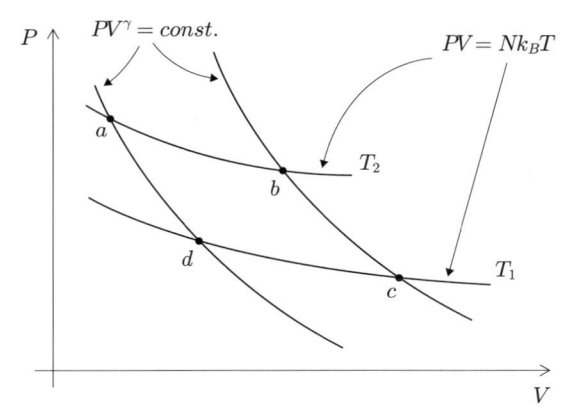

図 3.1 等温または断熱的なカーブの例を示す

$$TV^{\gamma-1} = constant \qquad (3.10)$$

という式と等価になる.

この式は 2 つの方程式

$$T_2 V_b^{\gamma-1} = T_1 V_c^{\gamma-1}$$
$$T_2 V_a^{\gamma-1} = T_1 V_d^{\gamma-1} \qquad (3.11)$$

を導く. この 2 つから,

$$\frac{V_b}{V_a} = \frac{V_c}{V_d} \qquad (3.12)$$

ということがわかる. これは, 式 (3.5) と式 (3.6) の Log で表示されたものと同じもので, つまり, 最後に

$$\frac{\Delta Q_1}{T_1} = \frac{\Delta Q_2}{T_2} \qquad (3.13)$$

という式が得られる. この式は, 系のエントロピーを表し, 2 体の $\Delta Q/T$ という値は, 可逆な機関であれば, 決して変化しない. 私たちはまた, 式 (3.13) を

$$\frac{\Delta Q_1}{T_1} = \frac{\Delta Q_2}{T_2} \qquad (3.13)$$

This is the equation expressing the *entropy* of a system. The value $\Delta Q/T$ of two bodies never changes in a reversible engine. We can also write eq (3.13) this way

$$\frac{\Delta Q_1}{T_1} = S = \frac{\Delta Q_2}{T_2} \qquad (3.14)$$

where S is called the entropy of the system. In a real case, of course, the entropy will increase, and will be given by this integral relation

$$S_b - S_a = \int_a^b \frac{dQ}{T} \qquad (3.15)$$

So another way to express the two laws of thermodynamics is as the following:

first law: the energy of the universe is constant

second law: the entropy of the universe is always increasing

Thermodynamics is a complex discipline that involve deep study and we do not want to go too much in details here, however please try to consider the philosophical implication of the few concepts that we have given here.

$$\frac{\Delta Q_1}{T_1} = S = \frac{\Delta Q_2}{T_2} \tag{3.14}$$

とも書けるが，このとき S というのは，系のエントロピーである．現実の場合，エントロピーは増大する．そして，

$$S_b - S_a = \int_a^b \frac{dQ}{T} \tag{3.15}$$

という関係になる．

つまり，熱力学の法則を別な言い方で言い表すと，

第 1 法則：系のエネルギーは常に一定である．

第 2 法則：系のエントロピーは常に増大する．

ということである．

熱力学は深く学ぶには複雑な分野だが，最初から詳細だけに捕われずに，そもそもの概念が何を意味するのかということを考えてほしい．

Chapter 4 Distributions

4.1 What is a "distribution"?

In thermodynamics we will use the term "distribution"[18] many times. A distribution is an abstract word that means the way a certain random variable organize in space, or in time, or in respect to any other variable. For example consider how tall is the people in your class. You can have many people around 160-170 cm, few taller ones and -say- three or four people below 160 cm. If you plot the number of people in certain range of height, you have the *distribution* of height in your class. This term is general, we can have of course the distribution of velocities of molecules in a gas, for example. This curve, the distribution, will represent how the velocities are distributed, or in other words, it will show you if some velocity is more common or frequent than other. Let's see in detail what this mean in the following paragraphs.

4.2 The distribution density in an air column

Let's consider a gas that is inside a big container. This gas is an ensemble of molecules, that are freely moving around. Each molecule has its own weight. This weight is definitely affecting how the molecules behave. The molecules at the bottom of the room will feel the pressure of all the molecules that are above.

What is the effect of the weight? Can we deduce a mathematical relation to describe the effect of the weight on the gas? To answer we can imagine to have two parallel invisible layer of area A that are separated by a very small

---

18) distribution: 分布.

第4章　分　　布

4.1　分布とは何だろう？

熱力学のなかで，私たちはよく分布という言葉を使う．分布というのは抽象的な言葉だが，その意味は「ランダムな変数を持つものが空間・時間・または他の変数によって，どのようにランダムに編成しているのかを示す方法」である．たとえば，あなたのクラスにいる人の身長について考えてみよう．おそらく，その身長はだいたい 160 cm から 170 cm ではないだろうか．数人の人はそれよりも高く，3，4 人は 160 cm より低いだろう．もしもあなたが，クラスの全員の，身長の高さとその人数をプロットすれば，あなたは「あなたのクラスのメンバーの身長分布」を調べたことになる．これは通常，例えば「気体分子の速度分布」のように使う．グラフでは，どんなふうに速度が分布しているのかを表している．他の言葉で言い換えると，どの速度がより一般的なのかを示している．では，その意味について詳しく考えてみよう．

4.2　気柱の中の空気の密度分布

大きなコンテナに入った気体について考えてみよう．この気体は自由に動き回る分子たちのアンサンブルである．それぞれの分子は質量を持っている．この質量は分子の振る舞いに影響する．コンテナの下のほうにある分子はコンテナの上にあるすべての分子の圧力を感じるだろう．

重さの影響はどのようになるだろうか．私たちは数学的な関係から，気体の重さの影響を導きだすことができるだろうか．そのために，とても小さい距離 dh だけ離れた面積 A の平行な 2 つの平面について考えよう．ひとつの面に働く微分の力は

$$dF = N < F > \tag{4.1}$$

4.2 The distribution density in an air column

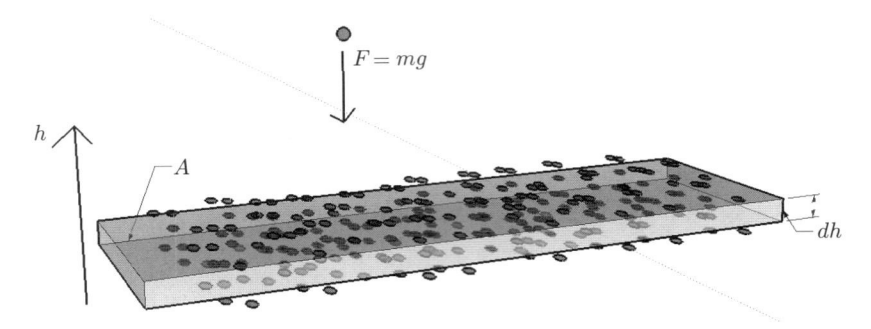

Fig. 4.1 The molecules of gas are indicated by dots. We consider the forces acting on two imaginary planes of area A. Each molecule weight is $F = mg$ and the distance between layer is an *infinitesimal* (dh).

infinitesimal distance dh. We consider that the *differential* force acting on a single layer is:

$$dF = N <F> \qquad (4.1)$$

where $<F>$ represents the average force of a single molecule and N the number of molecules enclosed in the box between the two layers (see figure 4.1). Please do not forget that this imaginary box is very small and its thickness is infinitesimal.

We know that the number of molecules can be calculated if we know the *molecular density* of the gas. This value is usually known and it is indicated by the letter n, $n = N/V$, where the volume V is in our case the area of the imaginary box: A multiplied by its infinitesimal thickness dh.

Now, the force exerted by a single molecule is just its weight, that is $<F> = -mg$. Let's put all these things together:

$$F = -mg$$
$$n = \frac{N}{Adh} \qquad (4.2)$$

If we substitute these in eq (4.1), we have

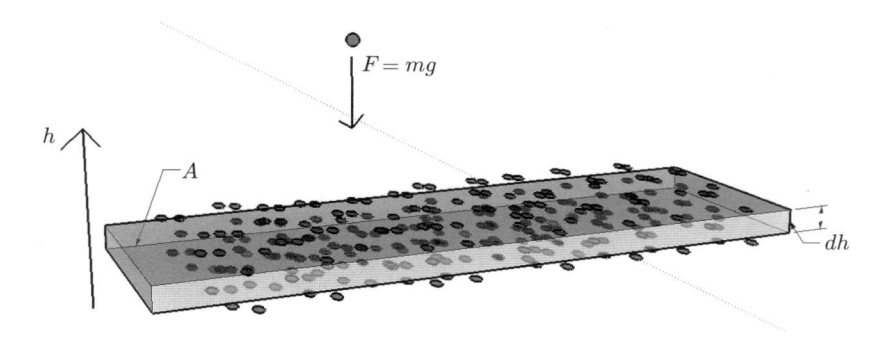

図 4.1 気体の分子はドットによって示されている．表面積 A をもつ平面に働く力を考える．それぞれの分子の力は $F = mg$ で，面は少しだけ（dh）離れている．

と書くことができる．ここで $<F>$ は 1 つの分子の力で，N は 2 つの平面に囲まれた箱に入っている分子の数である（図 4.1 参照）．ここで，この箱というのがとても小さくて，厚みがとても小さいということを忘れないでほしい．

気体の分子密度を知ることができれば，箱にいくつの分子が入っているかを計算することができる．この値は通常 n という記号で表し，箱の体積が V の時に $n = N/V$ となる．V は面積 A を厚さ dh 分足し合わせたものである．

1 つの分子によって及ぼされる力はその質量によって決定されるので，$<F> = -mg$ となる．まとめると，

$$F = -mg$$
$$n = \frac{N}{Adh} \tag{4.2}$$

となる．ここで式 (4.1) に代入すると，

$$dF = nAdh(-mg) \tag{4.3}$$

となる．今，圧力というのは単位面積あたりの力だということを思い出そう．そうすると，

$$dP = ndh(-mg) \tag{4.4}$$

と導ける．ここで，dP は 2 つの表面の圧力の差を表し，理想気体の場合には

4.2 The distribution density in an air column

$$dF = nAdh(-mg) \tag{4.3}$$

If we remember that the pressure is the force per unity of area, then we have:

$$dP = ndh(-mg) \tag{4.4}$$

Now, dP is the difference in pressure between the two surfaces, we know that for an *ideal* gas:

$$PV = NkT$$

$$P = nkT \tag{4.5}$$

We use the concept of *differentiation* and we obtain from the above eq (4.5)

$$dP = dnkT$$

We now substitute that on eq (4.4) and we have:

$$dnkT = ndh(-mg) \tag{4.6}$$

To integrate we simply put together the differential variables dn and dh:

$$\frac{dn}{n} = -\frac{mg}{kT}dh \tag{4.7}$$

Taking in account that kT and mg are constant values (they do not change with the height h), it is very easy to integrate this differential equation:

$$\int \frac{dn}{n} = -\frac{mg}{kT} \int dh \tag{4.8}$$

that results in

$$log\left(\frac{n}{n_0}\right) = -\frac{mg}{kT}\Delta h$$

applying the exponential function to both sides:

$$n = n_0 e^{-\frac{mg}{kT}\Delta h} \tag{4.9}$$

4.2 気柱の中の空気の密度分布

$$PV = NkT$$
$$P = nkT \tag{4.5}$$

ということを知っているので，式 (4.5) から

$$dP = dnkT$$

という関係を導くことができる．

この結果を式 (4.4) に代入すると，

$$dnkT = ndh(-mg) \tag{4.6}$$

となる．積分するために dn と dh の微分のパラメータでまとめると，

$$\frac{dn}{n} = -\frac{mg}{kT}dh \tag{4.7}$$

となる．kT と mg が（この2つのパラメータは高さ h で変わらないので）定数だということを考慮すると，この微分方程式を積分するのはとても簡単になる．

$$\int \frac{dn}{n} = -\frac{mg}{kT} \int dh \tag{4.8}$$

この結果を以下の式

$$log\left(\frac{n}{n_0}\right) = -\frac{mg}{kT}\Delta h$$

に当てはめて，両辺に指数関数をとると，

$$n = n_0 e^{-\frac{mg}{kT}\Delta h} \tag{4.9}$$

となる．

ここで n_0 は h_0 での密度 ($\Delta h = h - h_0$) を表す．この式が私たちに教えてくれることは何だろうか．まずはじめに，気体の密度は高さ h に依存して変化するということである．つまり，もしも大きな体積のものを考えるとしたら，高さによる気体の密度の違いを考慮する必要がある．式 (4.9) は密度が指数関数的に下がることを示している．つまり，下にある気体（$\Delta h \simeq 0$）は上

Here n_0 is the density at h_0 ($\Delta h = h - h_0$). What does this equation tells us? First of all, we understand that the gas density varies with the height h. So, if we have a very big volume, the density of the gas in it will change with height. Equation (4.9) shows an exponential decay, so the gas at the bottom (Δh near to zero) will have a higher density than gas at the top (Δh higher). If the gas is air, we immediately understand why if we are at sea level the air density is higher than up hill.

You will notice another thing: the exponential in eq (4.9) is modulated by the coefficient mg, the weight of the molecules[19]. We know that Oxygen (atomic mass 16) is heavier than Nitrogen (atomic mass 14), so if the air is a mixture of the two, equation (4.9) tells us that Oxygen should diminish earlier than Nitrogen. This is exactly what happens if we climb an high mountain. Not only all the air get thinner, but also the Oxygen drops faster than Nitrogen, so the relative ratio of Oxygen in the air drops.

Amazingly all these real life facts are deducible by the equation (4.9) that we just found by simple elementary calculations based on the ideal gas law!

Example 4.1

Lets calculate how much the ration Oxygen/Nitrogen in air changes with height. Suppose you are hiking uphill and measure the Oxygen/Nitrogen ratio at a starting point and this result to be R_1. When you climb 2000 meter and measure again obtain R_2. Use what learned here to deduce the variation of R of this Δh=2000 meters height difference.

We simply have to apply equation 4.9 several times. First of all, lets consider the variation of concentration of Oxygen for a rise of 2000

---

[19] weight というのは質量を意味するが，英語ではその質量にかかる力という意味でも使われる．たとえば，日本語でいうと重量キログラム（あるいはキログラム重）のような感じだろうか．

にある気体（Δh が大きいとき）よりも密度が高くなっている．もしもその気体が空気だとしたら，海抜ゼロにある空気の密度が，高い丘の上の空気よりも密度が濃い理由を理解することができる．

皆さんはもっと違うことにも気がついただろうか．それは，式 (4.9) の指数関数が，係数 mg によって影響を受けるということである．酸素（原子質量 16）は窒素（原子質量 14）よりも重いため，もしも空気が 2 種類の気体の混合物だとすると，式 (4.9) は，高さが高くなるほど酸素は窒素より早く減少するということを示している．これは私たちが山に登る時，実感できる．山登りの時は，ただ空気の密度が小さくなるだけではなく，窒素に比べて酸素の割合が減るために，空気中の酸素濃度が急激に落ちるのである．私たちの生活の中で体感する事象を，理想気体の方程式に基づいたシンプルで簡単な計算から得た式 (4.9) から導くことができるというのは，驚くべきことである．

例題 4.1

高度によって酸素と窒素の割合がどのくらい変わるか計算してみよう．ハイキングで山を，麓から酸素と窒素の割合を測りながら登るとしよう．このとき，麓での酸素/窒素の割合は R_1 であった．2000 m 登ったところで，再び測ると R_2 であった．この 2000 m の高さの差 Δh から，酸素と窒素の比 R をどうやって導けばよいだろうか．

答えは単純に式 (4.9) を何回か使えばよいだけである．まずはじめに，2000 m での酸素の濃度について考えると，

$$n^o = n_0^o e^{-\frac{m_o g}{kT}\Delta h} \tag{4.10}$$

となる．同様に窒素は

meters we have:

$$n^o = n_0^o e^{-\frac{m_o g}{kT}\Delta h} \qquad (4.10)$$

and for the Nitrogen at same hight we have:

$$n^N = n_0^N e^{-\frac{m_N g}{kT}\Delta h} \qquad (4.11)$$

here m_o and m_N represent the mass in grams of one atom of Oxygen and Nitrogen respectively. Similarly n^o represents the density of Oxygen atoms and so on. Now we divide the last two equation by each other and obtain:

$$\frac{n^o}{n^N} = \frac{n_0^o}{n_0^N}\left(exp\left(-\frac{m_o g}{kT}\Delta h + \frac{m_N g}{kT}\Delta h\right)\right) \qquad (4.12)$$

this is:

$$\frac{n^o}{n^N} = \frac{n_0^o}{n_0^N} e^{-\frac{g\Delta h}{kT}(m_o - m_N)} \qquad (4.13)$$

Now we simply have to rewrite our equation and put the correct numbers (we assume that the mass of an atom is its atomic weight divided the Avogadro number $N_A = 6 \times 10^{23}$, T = 300 K, g = 10 m/s^2, k = 1.38×10^{-23} and $m_o = 16$ g, $m_N = 14$ g)

$$R_2 = R_1 e^{-\frac{g\Delta h}{kT}\frac{(m_o - m_N)}{N_A}} \qquad (4.14)$$

$$R_2 = R_1 e^{-\frac{10 \times 2000}{300 \times 1.38 \times 10^{-23} \times 6 \times 10^{23}}(16-14) \times 10^{-3}} \qquad (4.15)$$

$$R_2 = R_1 e^{-0.0161} = 0.984 \times R_1 \qquad (4.16)$$

so the ration of Oxygen diminishes of about 1.6%, which is in the order of magnitude correct.

4.2 気柱の中の空気の密度分布

$$n^N = n_0^N e^{-\frac{m_N g}{kT}\Delta h} \tag{4.11}$$

となる.

ここで m_o と m_N は酸素と窒素の質量をそれぞれ表している. 同様に n^o は酸素の密度を示している. そこで, 上の 2 つの式から

$$\frac{n^o}{n^N} = \frac{n_0^o}{n_0^N}\left(exp\left(-\frac{m_o g}{kT}\Delta h + \frac{m_N g}{kT}\Delta h\right)\right) \tag{4.12}$$

を得る. つまり,

$$\frac{n^o}{n^N} = \frac{n_0^o}{n_0^N} e^{-\frac{g\Delta h}{kT}(m_o - m_N)} \tag{4.13}$$

ということである.

さて, 具体的な数字を入れて, 計算してみよう. 質量は原子質量をアボガドロ数 $N_A = 6 \times 10^{23}$ で割ったもの, $T = 300$ K, $g = 10$ m/s^2, $k = 1.38 \times 10^{-23}$ J/K, 1 mol あたりの質量は $m_o = 16$ g, $m_N = 14$ g である. すると

$$R_2 = R_1 e^{-\frac{g\Delta h}{kT}\frac{(m_o - m_N)}{N_A}} \tag{4.14}$$

$$R_2 = R_1 e^{-\frac{10 \times 2000}{300 \times 1.38 \times 10^{-23} \times 6 \times 10^{23}}(16-14) \times 10^{-3}} \tag{4.15}$$

$$R_2 = R_1 e^{-0.0161} = 0.984 \times R_1 \tag{4.16}$$

つまり, 酸素の割合は 1.6% ほど小さくなっており, この計算結果は桁の範囲では正しい.

4.3 The Boltzmann law

Now let's consider what we learned in the previous chapter in more general terms. Let's consider a force F between any particles of a gas. In the previous chapter the only force acting on the particles were the own particle weight. Now this force can be ANY force, for example energy potential between the gas particles, due to mutual attraction... or anything else. The gravity on the previous chapter example was acting only on the Z axis (the vertical direction). Instead, the more general force we choose here acts simultaneously on the three dimensions. For example: the attraction between molecules is a force that has no preferred direction. Any molecule feels a force coming from other nearby molecules on any direction in space.

However, now for simplicity we choose one arbitrary direction and we consider what is happening on our system of molecules along this single direction. Let's call this direction x and consider two parallel plane surfaces of area A between the gas. These surfaces are normal to the direction x as shown in figure 4.2.

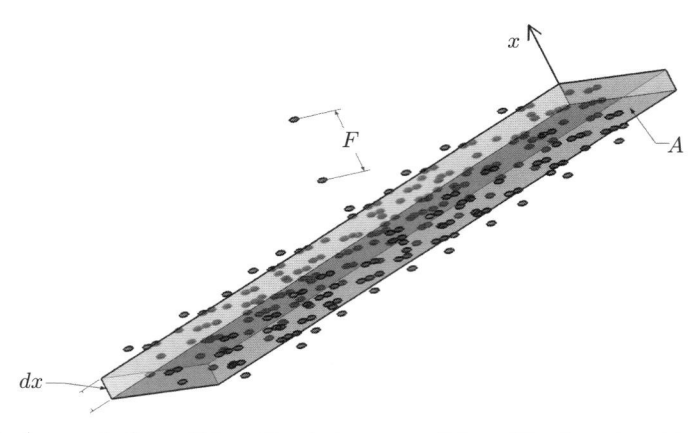

Fig. 4.2 A generic force F is acting between particles. The force is arbitrary and acts along all the directions in space, however we consider only the effects of the force along an arbitrary axis x as represented in this sketch.

4.3 ボルツマンの法則

ここでは前章のことについて，もっと一般的に考えよう．気体のすべての粒子間に働く力 F について考えてみよう．前の章では，分子に働く力はその粒子の質量によるものだけを考えた．ここでは，すべての力について考えよう．例えば分子間に働く相互引力作用や，その他の要因によるポテンシャルエネルギーなども考える．前の章でも取り上げた重力は，z 方向（垂直方向）にのみ働く．一般的に力は3次元的に作用する．例えば，分子間に働く相互作用には，特定の方向というものがない．分子は，それがどの方向でも，近くにきた分子に対して力を感じる．

しかし，簡単に考えるために，任意に1つの方向を選ぶ．そして，この1つの方向に沿って，分子がどう動くか考える．この任意の1つの方向を x とおき，面積 A をもつ平行な面の間に気体分子があると考える．この面は図 4.2 に示すように，x 方向に対して垂直である．

まさしく前の章で扱ったこれらの面に働く力の差 dF は，すべての分子の力の合計なので，その空間に存在する分子の個数を掛けて，

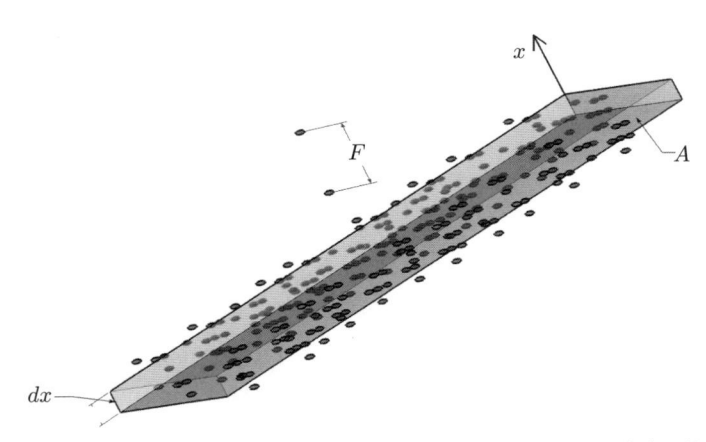

図 4.2 一般的な力 F は分子間に作用する．この力は任意で，すべての方向に働く．しかし，ここで私たちは1つの方向 x についてのみ考え，図に示す．

Exactly as we did on the previous chapter, the difference of force dF acting between these imaginary planes can be modelled as the net total force acting on every single particle, multiplied by the number of particles in the volume between the two surfaces. We can write this total force as:

$$dF = <F> n dV$$

where $<F>$ is the average force for a single gas molecule. The ndV is the gas density n ($= N/V$) multiplied by the volume between the two surfaces (again $dV = Adx$ where dx is the distance between them). So we have:

$$dF = <F> n dV$$
$$dF = <F> n A dx$$
$$dP = <F> n dx \qquad (4.17)$$

P is the difference in pressure between the two surfaces dF/A, we know that for an *ideal* gas:

$$PV = NkT$$
$$P = nkT \qquad (4.18)$$

With n we mean the *gas density* $n = N/V$, in other words: the number of molecules per unity of volume. Due to the variation of pressure the density varies. We use again the concept of *differentiation* and we have from the above eq (4.18)

$$dP = dn kT$$

Substituting in eq (4.17) we have:

$$<F> n dx = kT dn$$

This is an easy differential equation. We know that for any product $<F> dx$ corresponds an equal and opposite *work* $-dW$, we gather the relevant variables dx and dn:

4.3 ボルツマンの法則

$$dF = <F> ndV$$

と書くことができる．ここで $<F>$ は1つの分子に働く平均の力である．ndV は気体の密度 $n\,(=N/V)$ と2つの平面で挟まれた体積を掛けたものである（再び，$dV = Adx$ と書いておく．dx は2つの平面間の距離）．つまり，以下のようになる．

$$dF = <F> ndV$$
$$dF = <F> nAdx$$
$$dP = <F> ndx \tag{4.17}$$

P は2つの表面によって決まる圧力差 dF/A で，理想気体の場合には

$$PV = NkT$$
$$P = nkT \tag{4.18}$$

となる．n は気体の密度 $n = N/V$ である．気体の密度というのを別の言葉で言うと，単位体積あたりの分子の数である．圧力が変化すると，密度も変化する．ここで再び微分のコンセプトを用いると上の式 (4.18) は

$$dP = dnkT$$

となる．

式 (4.17) に代入すると，私たちは

$$<F> ndx = kT dn$$

という式を得ることができる．これは簡単な微分の式である．$<F> dx$ という積は，同じ大きさで逆向きの仕事 $-dW$ に一致する．私たちは dx や dn といった関係のある変数について次式を得る．

$$\underbrace{<F> dx}_{-dW} = kT \frac{dn}{n}$$

$$\overbrace{<F> dx}^{-dW} = kT \frac{dn}{n}$$

then we take the integral, we remember that kT are constants:

$$-\int dW = kT \int \frac{dn}{n}$$

$$-\Delta W = kT \log\left(\frac{n}{n_0}\right)$$

$$\frac{-\Delta W}{kT} = \log\left(\frac{n}{n_0}\right)$$

we name $\Delta W = W$, then apply the exponential function to left and right of the equation and we have:

$$\frac{n}{n_0} = e^{-\frac{W}{kT}}$$

$$n = n_0 \, e^{-W/kT} \qquad (4.19)$$

this is the dependence of the gas density, for **any potential force** acting on the molecules. The generality of this relation, that is usually called *Boltzmann law*, is something awesome. It says that the gas density of an ideal gas is related in a negative exponential dependence with the potential energy relative to the forces acting on these particles. Please remember that the relation we obtained above is correct even if the force F is a realistic force that acts on the three dimensions. Eq (4.19) limits the dependence of the density n along the one dimensional direction x. However, the force F that causes the dependence can be a general force acting on all three dimensions.

Now let's look at the last generalization: can we represent the gas density $n = N/V$ in another way? Clearly, n it is something proportional to the **probability to find a gas molecule** in a given volume V. The more molecules in a volume, the higher probability to find a molecule, less

4.3 ボルツマンの法則

kT は定数だということを思い出すと以下のようになる．

$$-\int dW = kT \int \frac{dn}{n}$$

$$-\Delta W = kT log\left(\frac{n}{n_0}\right)$$

$$\frac{-\Delta W}{kT} = log\left(\frac{n}{n_0}\right)$$

ここで $\Delta W = W$ だということを仮定すると，指数関数を用いて，

$$\frac{n}{n_0} = e^{-\frac{W}{kT}}$$

$$n = n_0 \, e^{-W/kT} \tag{4.19}$$

となる．これは気体密度のポテンシャル依存性を表しており，どんなポテンシャルでも成り立つ．通常ボルツマンの法則と呼ばれるこの関係の一般性は，畏敬の念を抱かせるものである．この式が何を意味しているのかというと，理想気体の密度が，その分子に働く力のポテンシャルエネルギーの負の指数関数で表されるということである．上で導いた関係は，力 F が三次元的に働く場合でも成り立つ，ということを覚えておいて欲しい．式 (4.19) は一次元の x 方向に沿った密度 n に限定しているが，力 F は一般に三次元に働く力である．

最後に一般化してみよう．気体の密度 $n = N/V$ について，他の方法で表現することはできるだろうか．n が一定の体積 V において気体分子を見つける確率に比例していることは明らかである．同じ体積中で分子の数が増えると分子を見つける確率が高くなり，数が減少するとその確率も下がる．そこで，もしもボルツマンの法則について書き直すことができるとしたら，以下のようになる．

$$f = const \, e^{-P.E./kT} \tag{4.20}$$

ここで f は気体分子を見つける確率で，P. E はポテンシャルエネルギーを表し，$const$ は任意の定数である．多くの本の中で，ボルツマンの法則というのはこの一般的な形式で書かれている．つまり式 (4.20) は，ある物理量の分布

molecules, lower probability. So, if we like, we can rewrite the Boltzmann law like this:

$$f = const\, e^{-P.E./kT} \qquad (4.20)$$

where f is the *distribution* probability of finding a gas particle, P.E. represents a general *potential energy* and *const* is any constant. In many textbooks the Boltzmann law is written in this more general form. You can read eq (4.20) as: *the distribution of a certain physical property f decays exponentially with speed proportional to the energy potential relating the physical property and inversely proportional to the system temperature.* Amazingly enough, equation (4.20) is applicable in a multitude of real physical systems!

Example 4.2

Suppose we have a molecular aggregation in which the molecules are kept together by a force that obey the Hooke law $F = -Kdx$, where dx is the intermolecular distance. You can image a system of molecules kept together by springs. Of course, these springs will never be quiet. They will continuously move a little around an equilibrium point. The intensity of movement is the temperature of the system. Using what we learned up to now, can we calculate the distance distribution of these molecules at a certain temperature?

Yes we can do that! First of all, we have to clarify what we mean with *distance distribution*. If we take a snapshot of the system, and we count the molecules at distance dx_1 and those at dx_2, dx_3 and so on, and we plot these numbers, this histogram is the *distribution* we mean. If so, we just use our Boltzmann distribution equation.

Suppose the Hooke constant is $K_h = 10^{-5}$ N/m, dx_1=10 nm, dx_2=14 nm our problem is to find the molecular density ratio n_1/n_2 at room temperature $T = 300$ Kelvin. We do this simply applying twice the

の確率 f は指数関数的に減衰し，減衰の速さは，物理的性質を表すポテンシャルエネルギーに比例し，系の温度に反比例すると解釈できる．驚くべきことに式 (4.20) は多くの実際の物理の系において適応できる．

> **例題 4.2**
>
> フックの法則 $F = -Kdx$ に従う力で繋がっている分子の集団について考えよう．分子間距離は dx である．分子はバネで繋がっていると想像しよう．もちろん，バネはじっとしていない．バネは絶えず少しずつ平衡位置の付近を動いている．この動きの大きさは，系の温度と同じである．これまで学んで来たことから，ある温度での分子の距離の分布について考えてみよう．
>
> 大丈夫，できるはずである．まずはじめに，分子の距離の分布とは何かを明確にしておく必要がある．もしもこの系の瞬間写真を撮ることができたら，分子の距離 dx_1, dx_2, dx_3, , ，を測定し，距離とその分子数を数えることができる．このヒストグラムが分布ということである．もしもこの方法が可能であれば，ボルツマン分布の式を使ってすぐに求めることができる．
>
> バネ定数が $K_h = 10^{-5}$ N/m, $dx_1 = 10$ nm, $dx_2 = 14$ nm の場合，300 K での分子密度の比 n_1/n_2 を考えてみよう．ごく単純に，ボルツマン方程式を 2 回適応すればよい．すると

Boltzmann relation:

$$n_1 = n_o \times e^{-P.E/kT}$$
$$n_2 = n_o \times e^{-P.E/kT} \qquad (4.21)$$

Since the potential energy for Hooke's law is $P.E = \frac{1}{2}K_h dx^2$, then

$$n_1 = n_o \times e^{-K_h dx_1^2/2kT}$$
$$n_2 = n_o \times e^{-K_h dx_2^2/2kT} \qquad (4.22)$$

if we divide these two equation we have the relation we are looking for

$$\frac{n_1}{n_2} = \frac{e^{-K_h dx_1^2/2kT}}{e^{-K_h dx_2^2/2kT}}$$
$$\frac{n_1}{n_2} = e^{-K_h(dx_1^2 - dx_2^2)/2kT} = e^{-K_h(dx_1+dx_2)(dx_1-dx_2)/2kT} \qquad (4.23)$$

Substituting[20] our numbers we have:

$$\frac{n_1}{n_2} = e^{-10^{-5}(10+14)\times 10^{-9}(10-14)\times 10^{-9}/(2\times 1.3810^{-23}\times 300)} = e^{0.11} \approx 110\% \qquad (4.24)$$

this means that at anyplace in space, at any moment in time, the number of molecules separated by $dx_1 = 10$ nm is about 10% more than those separated by $dx_2 = 14$ nm. This is logically what you would expect. We need more energy to separate molecules if they are kept together by a spring! So the further distance we look from, the less molecules we will find.

4.4 The concept of distribution of a general potential

The force acting on a particle is in general the derivative of a potential energy. For every force there must be a potential energy that generates this

---

20) substitute: 代入する.

$$n_1 = n_o \times e^{-P.E/kT}$$
$$n_2 = n_o \times e^{-P.E/kT} \tag{4.21}$$

となり，フックの法則のポテンシャルエネルギーは $P.E = \frac{1}{2}K_h dx^2$ なので，

$$n_1 = n_o \times e^{-K_h dx_1^2/2kT}$$
$$n_2 = n_o \times e^{-K_h dx_2^2/2kT} \tag{4.22}$$

となる．2つの式を分けると，

$$\frac{n_1}{n_2} = \frac{e^{-K_h dx_1^2/2kT}}{e^{-K_h dx_2^2/2kT}}$$
$$\frac{n_1}{n_2} = e^{-K_h(dx_1^2-dx_2^2)/2kT} = e^{-K_h(dx_1+dx_2)(dx_1-dx_2)/2kT} \tag{4.23}$$

となる．実際に値を入れてみると，

$$\frac{n_1}{n_2} = e^{-10^{-5}(10+14)\times 10^{-9}(10-14)\times 10^{-9}/(2\times 1.3810^{-23}\times 300)} = e^{0.11} \approx 110\% \tag{4.24}$$

となる．この結果は，どんな場所でも，どんな瞬間でも，$dx_1 = 10$ nm だけ離れた分子は $dx_2 = 14$ nm のものより 10% 多く存在するということを示している．これは，知りたかった情報である．このバネで繋がった分子の場合，離れるほど多くのエネルギーが必要なので，離れた距離にいる分子の数は少ないということがわかる．

4.4 一般的なポテンシャルの分布についての概念

分子に働く力は，一般的にはポテンシャルの導関数である．どの力についても，その力を生み出すポテンシャルエネルギーというものが存在しなくてはならない．力はそのポテンシャルの微分から導かれ，かつマイナスの符号がついていて，ポテンシャルの勾配が常に力と逆向きであるということを示してい

force. The force is the derivative of the potential, multiplied with a minus sign to indicate that the potential is always opposite to the force. So if we know the mathematical expression of the potential we can calculate the expression of the force. This is one of the great principles of nature that is at the basis of physics.

In the example of the air column that we discussed above, the potential energy is

$$dW = -(mg)dh$$

where dh is an infinitesimal variation of height, m the average particle mass and g the gravitational constant $g = 9.81$ m/s^2. In this case the force F is then

$$F = -\frac{dW}{dh} = mg$$

This is true in the simplest case of an air column where molecules are subject only to the force of their own weight. But what happens if we consider a more complex case? For example let's suppose that each molecule has a slight attraction to the next one. This is actually true. Even gas molecules, if they are placed very near each other they feel a small force of attraction. That's why if we remove the kinetic energy (reduce temperature) the gas tends to transform to a liquid. Then, let's ask ourselves: what is usually the potential energy responsible of the forces between two molecules of gas? Of course we do not know this answer exactly. However, we can firstly consider a very simple case: two masses (our molecules) connected by an ideal spring. We choose this, because we know how a spring works. The system we imagine looks like a simple bi-atomic molecule...!

We know that in the *Hooke's law*, a spring gives a force $F = -kx$ where x is the extension of the spring respect to a *rest* position x_0. In our case if we call the distance connecting the two molecules as R and the rest position as

4.4 一般的なポテンシャルの分布についての概念

る．もしもポテンシャルの数学的な式を知っていたら，私たちは力の表現を計算によって得ることができる．これは，物理学の基本的な素晴らしい原理の1つである．

これまでの章で取り扱って私たちが議論してきた気体分子について考えると，ポテンシャルエネルギーは

$$dW = -(mg)dh$$

であった．dh は高さに関する微小変化量で，m は気体分子の平均質量，g は重力加速度（$g = 9.81 \,\mathrm{m/s^2}$）を表す．この場合，$F$ というのは，

$$F = -\frac{dW}{dh} = mg$$

となる．

4.2 節では，気体分子がその重量によってしか力を受けない場合を考えたので，この結果は当てはまった．しかし，もっと複雑な場合にはどうなるだろうか．例えば，隣り合う分子から微小な引力を感じるとしたらどうなるだろうか．現実の場合はそのようになっていて，気体分子でさえも，もしそれがとても近くに位置していたら，小さい引力を感じる．このため，運動エネルギーが小さくなると気体の温度が低下し，液体になろうとする．ではここで，自分自身に問い直そう．2つの気体分子の間に働く力を表すポテンシャルエネルギーとは何だろうか？　私たちはそのものズバリの解答は知らない．しかし，最初に私たちの分子を2つの質量が理想的なバネででつながっているというモデルとして単純に考えることはできる．これを選んだのは，もちろん，バネがどう動くかを知っているからである．この系は単純な2原子分子のモデルだと考えることができる．

私たちはフックの法則を知っているので，バネが及ぼす力は $F = -kx$ であるとわかる．ここで x はバネの静止位置 x_0 に対する伸びを示す．私たちが考えている状況では，2つの原子が R という距離にあり，その静止位置が r_0 であるとすると，2つの原子に働く力は

r_0, then the force between them is

$$F = -k(R - r_0)$$

Let's remember that in general $dW = -Fds$ where ds is the displacement along the direction of the force. Then, we multiply our F by a small displacement dR, we obtain the infinitesimal potential energy that generate the force

$$dW = FdR = k(R - r_0)dR$$

As we said, if we integrate this we obtain the potential energy (P.E.):

$$P.E. = \frac{1}{2}R^2k - r_0Rk \qquad (4.25)$$

We can plot this potential and its relative force, see figure 4.3.

This was an ideal case, however, in reality the potential between two gas molecules usually has a shape similar to that in figure 4.4 (near: repulsion, far: attraction, very far: no-effect)[21]. In a gas system we do not have only two molecules, but many of them. So using the same integration methods, it is possible to derive the total potential energy W just summing up all the contributions, like this:

$$W = \sum_{ij} V_{ij} \qquad (4.26)$$

Now let's take into consideration the general Boltzmann law. From eq (4.26) the probability to find a molecule will be:

$$f = const. \times e^{-\sum_{ij} V_{ij}/kT} \qquad (4.27)$$

What will happen if the temperature changes? If the temperature decreases to low values then the term $W = \sum_{ij} V_{ij}$ will be dominant. So the probability of finding particles at the minimum potential r_0 will be higher

---

21) repulsion: 斥力, attraction: 引力.

$$F = -k(R - r_0)$$

である.

さて，一般的な法則 $dW = -Fds$ を思い出そう．ここで ds というのは力の方向の変位のことである．つまり，先程の F を小さな変位 dR と掛け合わせると，私たちは微小のポテンシャルエネルギー

$$dW = FdR = k(R - r_0)dR$$

を導くことができる．前に言ったように，積分すると，

$$P.E. = \frac{1}{2}R^2 k - r_0 R k \tag{4.25}$$

とポテンシャルエネルギー ($P.E.$) を得ることができる．私たちはポテンシャルとそれによって派生した力をプロットすることができ，その関係を示したものが図 4.3 である.

これはもちろん，理想的な場合についてのものである．しかし，現実の場合においても，2 つの気体分子の間に作用するポテンシャルは，図 4.4 に示したものに似た形になる（近傍では反発力，少し遠くになると引力，もっと遠くになると作用しなくなる）．気体の系では，分子は 2 つだけではない．もっとたくさんある．そこで，同じような積分のやり方を使うと，すべての寄与について

$$W = \sum_{ij} V_{ij} \tag{4.26}$$

のように足し合わせるだけで，全ポテンシャルエネルギー W を導くことができる.

さて，ここで，一般的なボルツマンの法則を用いて考察をしてみよう．式 (4.26) より，分子を見いだす確率は

$$f = const. \times e^{-\sum_{ij} V_{ij}/kT} \tag{4.27}$$

である．温度が変化するとどうなるだろうか．もし温度が低くなると，$W = \sum_{ij} V_{ij}$ が支配的になる．つまり，一番小さなポテンシャル r_0 の位置で分子

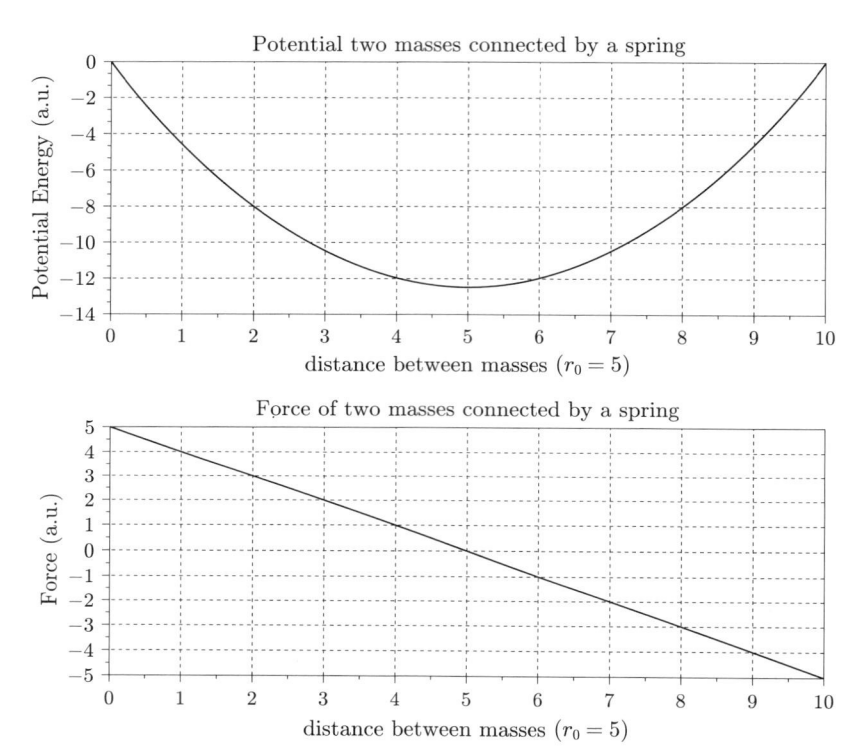

Fig. 4.3 The potential of an ideal force connecting two molecules: a perfectly elastic spring. On the top we see the potential, at the bottom the force.

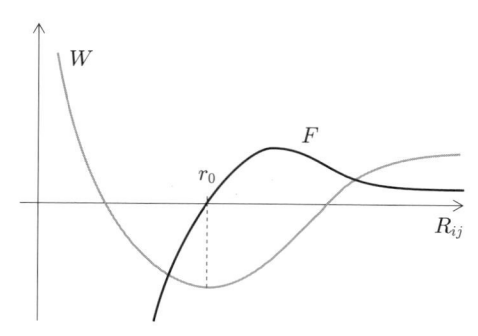

Fig. 4.4 A schematic representation of the potential shape of two molecules attracting each other (gray line) and the corresponding force (black line).

4.4 一般的なポテンシャルの分布についての概念

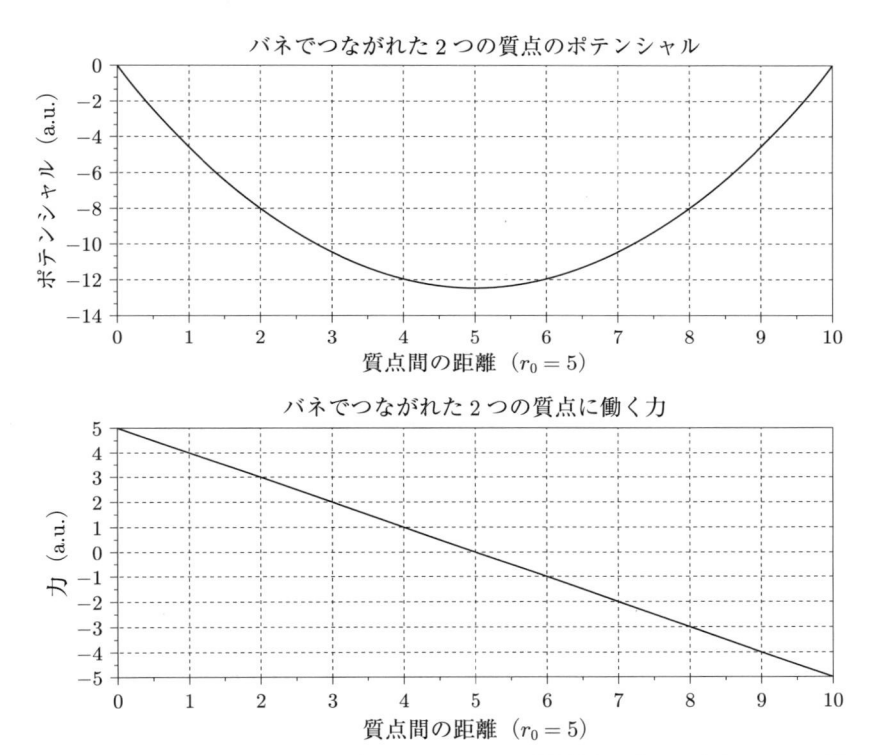

図 4.3 理想的な 2 つの分子間に働くポテンシャル．理想的なバネでつながっている場合について．上図はポテンシャルを示し，下図は力について示したもの．

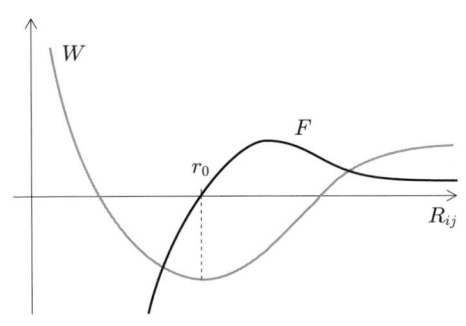

図 4.4 2 つの分子間に働くポテンシャル（グレー）と力（黒）を模式的に表したもの．

(because in that point W is minimum, so the exponential is maximum).

On the contrary, if the temperature increases, and get higher than the total potential $W = \sum_{ij} V_{ij}$, the exponent gets close to zero for any value of R, even the minimum point r_0 will have little influence. The exponential function becomes nearly one for every potential. This means that the particles will be randomly distributed (same probability to find them at any distance).

What we have just described in very rough terms is the phenomena of *evaporation*. If the temperature is low, majority of particles will be located at a rest distance from the others (like in a solid or a liquid), if temperature increases, the particles will fly away and they will be randomly separated, exactly what we observe in a gas. Equation 4.26, despite its simplicity and the basic assumptions we used to derive it, it is able to explain well the phenomenon of evaporation of materials with temperature. This is already a good result, but we achieve it without knowing the real shape of the potential W. Let's try to go further and assume to know the mathematical expression of the potential.

Example 4.3

Suppose that the force that is keeping two molecules together is linear and centered in $x_o = 5$, so it has the shape of figure 4.5 and it is expressed by this formula:

$$F = -(x - x_o)$$

(A negative force is attractive and a positive repulsive)
Can you find the shape of the potential of this force?
We have to consider that Force and the potential are related by

$$F = -dW/dx,$$

を見いだす確率が大きくなるということである．なぜなら，その点では W は最小で，つまり，指数関数が最大になる点だからである．

反対に，温度が上昇し全ポテンシャルエネルギー $W = \sum_{ij} V_{ij}$ よりも高くなった場合，指数はどの R に対してもゼロに近づき，最小点 r_0 においても，ほとんど影響を受けない．温度が高いと指数関数は（そのポテンシャルの値がどうであれ）1 に収束する．これが意味することは，粒子はランダムに動き回っており，すべての距離において粒子を見いだす確率は等しくなるということである．

これは，**蒸発**の現象について非常に大雑把に記述したものである．もしも温度が下がると，大多数の粒子は（固体や液体などと同様に）他の粒子に対して一定の距離に位置し，温度が上がると粒子は飛び回り粒子間の距離はランダムになる（まさに気体の場合）．式 (4.26) で簡単化するために仮定をしたが，温度の上昇によって蒸発するという現象を説明することができた．この結果は現実によくあったが，私たちはポテンシャル W の本当の形を知らないままに得た結果である．そこで，この先，ポテンシャルの数学的な表現についてもっと詳しく調べてみたい．

例題 4.3

2 つの分子をつなぐ力が線形で，中心が $x_o = 5$ のときを考えよう．力は図 4.5 のようになっていて，$F = -(x - x_o)$ という式で表すことができる（負の力は引力で，正は斥力）．この力のポテンシャルを描いてみよう．

力とポテンシャルの関係は以下のようになっている．

$$F = -dW/dx,$$

つまり $dW = -Fdx$ となり，$W = \int (x - x_o)dx = 1/2 x^2 - 5x + const$ となる．これを描くと，図 4.6 に示すような下に凸な放物線になる．

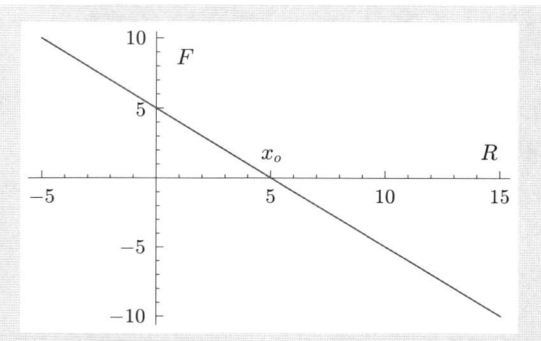

4.4 The concept of distribution of a general potential

Fig. 4.5 The force in the example is zero in $x_o = 5$. If two particles are placed at distance x and subject to this force, they repel if $x < x_o$ and attract if $x > x_o$.

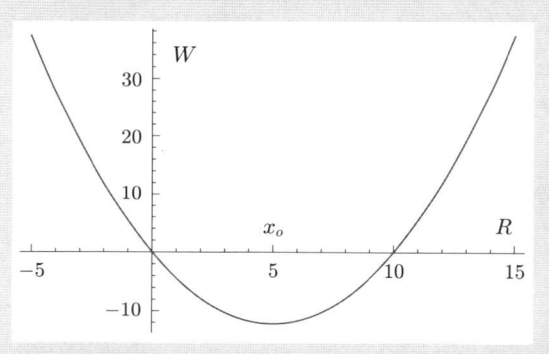

Fig. 4.6 The minimum of the potential is the point $x_o = 5$, where the force is zero. It is a *stable* point of the system because the concavity of the curve is positive.

so $dW = -Fdx$ and $W = \int (x - x_o)dx = 1/2 x^2 - 5x + const.$

The resulting shape of this curve is a parabola with positive concavity, as plotted in figure 4.6[22].

---

22) 実際にやってみることが大切！

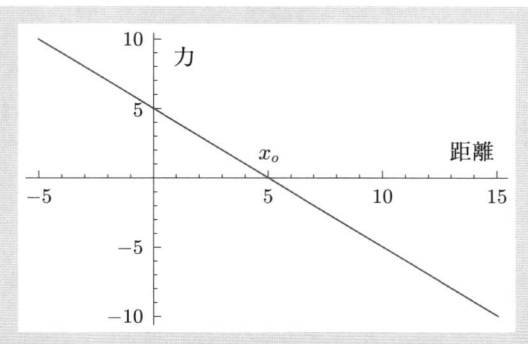

図 4.5 力と分子の距離の関係を示した図. $x_o = 5$ で力はゼロになる. $x < x_o$ では斥力, $x > x_o$ では引力を受ける.

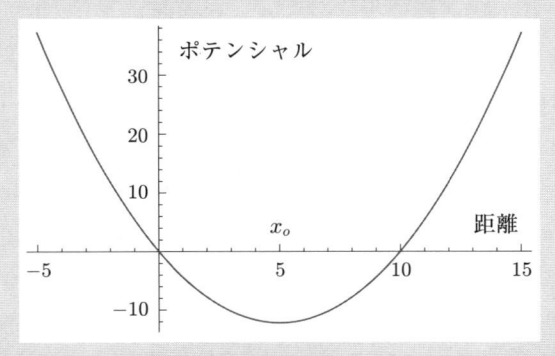

図 4.6 ポテンシャルと分子間距離の関係. ポテンシャルが最も小さくなるのは $x_o = 5$ の時で, これは力がゼロになるところと一致する. ポテンシャルの放物線が下に凸なので, $x_o = 5$ で系が安定になる.

Chapter 5 Various phenomena explained by thermodynamics

5.1 Physical states: gas, liquid and solids

Let's consider again the hypothetical situation in which molecules are connected by a simple spring. In this case, as we said above, the potential is derived by the Hooke's law as in equation (4.25). Let's substitute it in the Boltzmann formula, we obtain the Boltzmann distribution:

$$f = const. \times e^{-(\frac{1}{2}R^2\bar{k} - r_0 R\bar{k})/kT} \tag{5.1}$$

The symbol $\bar{k}$ is used to distinguish the Hooke constant from the Boltzmann one. This relation tells us the probability of finding a molecule in function of the molecular distance. We use a computer and make a graph from eq (5.1). The simulation is done in -so called- *arbitrary units*[23], putting the *const.*, $\bar{k}$ and k values to one. Even if these parameters loose their real meaning, the distribution shape does not change. Results are shown in figure 5.1. Let's try to understand its hidden physical meaning: first of all remember that in this plot there is not explicit dependence on the position of the molecules. Only their relative distance is considered here. So our graph deals about something that can be located anywhere in space.

On the horizontal axis[24] we have the distance between molecules R. In the case of our "Hooke's law" potential, the minimum potential is realized at the *resting position*, $R = r_0 = 5$ in our simulation. Let's concentrate first

---

23) arbitrary units: 任意単位. 実験を行う際に, 実験のセッティングや測定機器の違いによって, 測定できる信号などの強度が違うことがしばしばある. そのようなときは単位は「任意単位」を用いる. その実験の中では（セッティングや測定系を変えなければ）比較できる場合もあるが, 他の実験結果とは比べることができない.

24) horizontal axis: グラフの横軸のこと. 縦軸は vertical axis という.

第5章 熱力学から考える様々な現象

5.1 気相・液相・固相という物理的な相について

さて，分子が単純なバネでつながったという仮定の状況について考えてみよう．この場合，前に述べた通り，ポテンシャルはフックの法則（式 (4.25)）によって得ることができる．これをボルツマンの式に代入してみよう．そうすると，ボルツマン分布

$$f = const. \times e^{-(\frac{1}{2}R^2\bar{k} - r_0 R\bar{k})/kT} \tag{5.1}$$

を得ることができる．

$\bar{k}$ はボルツマン定数 k ではなくバネ定数である．この関係は私たちに，分子を見いだす確率が分子間距離の関数であると教えてくれる．コンピュータを使って式 (5.1) のグラフを描こう．シミュレーションはいわゆる，任意単位において $Const$, $\bar{k}$, k の値をそれぞれ 1 として行った．もしもそのパラメータが実際の値と異なったとしても，分布の形は変わらない．結果を図 5.1 に示す．この結果に隠された物理的意味を考えよう．まず最初に，このプロットは原子の位置そのものを示しているわけではないことに注意しよう．このグラフより明らかになるのは，相互の距離についてだけである．つまり，このグラフは空間のどこかに位置する分子の相互の距離を示す結果である．

横軸が示しているのは分子間距離 R である．私たちが考えた「フックの法則」のポテンシャルの場合，ポテンシャルが最も小さくなる場所が「自然長」であり，シミュレーションでは $R = r_0 = 5$ として計算した．では次に，次の「熱い」曲線について考えてみよう．これは温度 T が（$T = 1000$）という高い温度での状況をプロットした曲線である．この曲線は，分子間距離について着目すると，どの距離の分子も統計的に同じ確率で存在しているということを

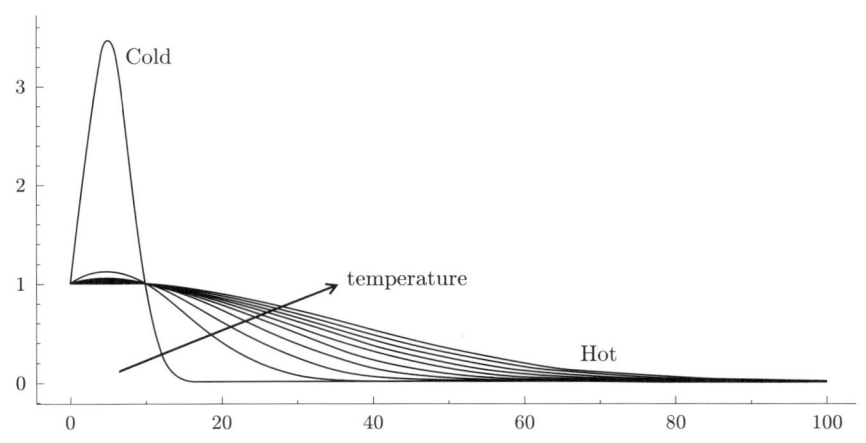

Fig. 5.1 In this simulation of eq. (5.1) we run temperatures $T = 10$ to $T = 1000$ (direction of temperature is indicated by the dark arrow "Temp") and set the resting position to $r_0 = 5$. Clearly, as the temperature increases the potential gets less important, because the function tends to a constant value everywhere. All molecules tend to be sparse and randomly placed far from the minimum (gas state). If temperature is low compared to the potential, the max probability to find the particles is when they are at the minimum of potential, exponential is at maximum. This can be the case of a liquid or a solid.

on the "Hot" curve, that has been plotted for high values of the parameter T ($T = 1000$). For this curve we can say that: at *any give point* in the gas, if we look at the distance between molecules (also called *intermolecular distance*[25]), statistically we see that they have all similar probability to exist. Instead, if we are at lower temperatures, we see that the curve has a maximum. And this maximum is exactly at the resting position $r_0 = 5$! The curve goes also to zero very quickly. The physical meaning is: intermolecular distances near r_0 are very probable, whereas different ones are not.

Let's repeat that this specific *distribution* is valid in any given point in space. In any point of space of our gas, we find the same distribution of

---

25) intermolecular distance: 分子間距離.

5.1 気相・液相・固相という物理的な相について

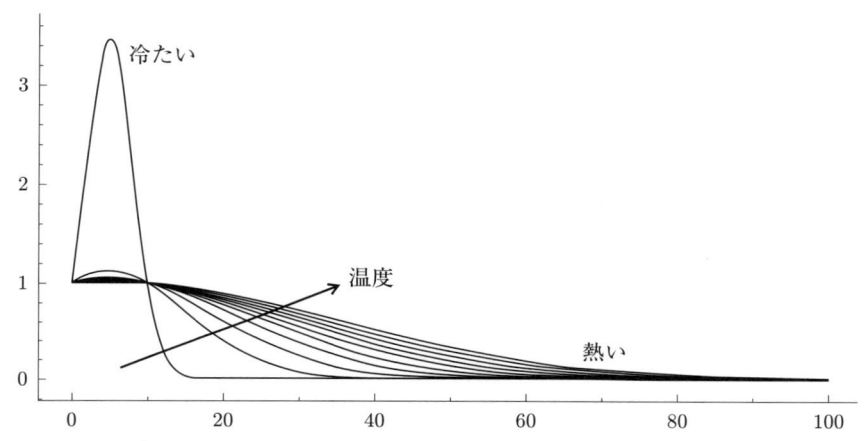

図 5.1 この式 (5.1) のシミュレーションにおいて，私たちは温度を $T = 10$ から $T = 1000$ まで計算を行った．温度の静止位置は $r_0 = 5$．明らかに温度があがると，ポテンシャルの重要性は失われる．なぜなら，関数はどこも一定値になる傾向があるからである．すべての分子は，バラバラになって一番小さい位置からランダムに位置する（気体状態）．もしも温度がポテンシャルに比べて低い場合には，粒子はポテンシャルの一番低い場所で見つけることができるだろう．これがつまり，液相や固相の場合である．

示している．言い換えると，「冷たい」温度では，曲線は最大値をもっている．この最大値というのは，まさに安定位置の $r_0 = 5$ である．そしてその曲線は急激にゼロに下がる．この物理的な意味は，ほとんどが分子間距離が短い分子として存在し，距離が遠い分子はほぼ存在しないということである．

ただし，空間的な分布は場所によって変わらないことを繰り返し述べておく．私たちは，気体のすべての点において，同じ分布を発見する．違う言葉でいうと，すべての気体分子において，分子間距離は図 5.1 に示すようにボルツマンの法則に則って存在する．覚えておいてほしいことは，分布の形はいつも同じであるが，（式 (4.27) 参照），その値は温度に依存して変化し，それが現実の世界での物理的な状況を（気相・液相・固体でさえも）表しているということである．

potential, in other words in any point of the gas the intermolecular distances have a probability of existing distributed as the Boltzmann law in figure 5.1. Remember: the *distribution* has always the same shape (equation 4.27), but the values of this distribution are different depending on temperature values and they represent different physical situations in the real world (a gas, a liquid or even a solid).

Example 5.1

Calculate the average intermolecular distance for molecules subject to a force $F = K_h(R - r_o)$, with $r_o = 1$ nm and at a temperature of $T = 13°C$. This problem is tricky... we know that the intermolecular distance change in time because of temperature, and we know also the statistical distribution of these continuous changing distances, but how we can determine the average at a certain temperature?

Let's start from the distribution we know, equation 5.1

$$f = const. \times e^{-(\frac{1}{2}R^2\bar{k} - r_o R\bar{k})/kT}$$

The intermolecular distance here is the variable R. We can assume that the average position of our molecules is coincident with the peak of the distribution, to determine this analytically we can calculate the derivative of it and put it equal to zero. In mathematical terms:

$$\frac{df}{dR} = 0 \qquad (5.2)$$

So, lets calculate this derivative:

$$\frac{df}{dR} = const. \times \frac{d}{dR}(e^{-(\frac{1}{2}R^2\bar{k} - r_o R\bar{k})/kT}) = 0 \qquad (5.3)$$

$$\frac{df}{dR} = const. \times e^{-(\frac{1}{2}R^2\bar{k} - r_o R\bar{k})/kT} \times \left[-\frac{1}{kT}(\bar{k}R - \bar{k}r_o)\right] = 0 \qquad (5.4)$$

since the exponential goes to zero only at infinite and the constant

5.1　気相・液相・固相という物理的な相について

例題　5.1

ポテンシャル $V = K_h(R - r_o)$ ($r_o = 1$ nm, $T = 13$℃) の平均の分子間距離を計算してみよう．この問題にはちょっとうまい作戦をとる必要がある．なぜなら温度によって分子間距離は変わるからである．温度による距離の統計的な分布はわかるが，どうやって「ある温度」での平均分子間距離を決めたらよいのだろうか．

まず，私たちが知っている式 (5.1) からはじめよう．

$$f = const. \times e^{-(\frac{1}{2}R^2\bar{k} - r_o R \bar{k})/kT}$$

分子間距離の変数は R である．平均的な分子間距離は，距離分布のピークであるとして考えよう．ピークの位置は導関数 = 0 から求められる．つまり

$$\frac{df}{dR} = 0 \tag{5.2}$$

である．実際に微分して考えてみると，

$$\frac{df}{dR} = const. \times \frac{d}{dR}(e^{-(\frac{1}{2}R^2\bar{k} - r_o R \bar{k})/kT}) = 0 \tag{5.3}$$

$$\frac{df}{dR} = const. \times e^{-(\frac{1}{2}R^2\bar{k} - r_o R \bar{k})/kT} \times \left[-\frac{1}{kT}(\bar{k}R - \bar{k}r_o)\right] = 0 \tag{5.4}$$

となる．指数関数は無限大のときのみゼロになるので，定数 $const.$ は消去でき，とてもシンプルな式

$const.$ is eliminated we have finally the very simple relation

$$\left[-\frac{1}{kT}(\bar{k}R - \bar{k}r_o)\right] = 0 \qquad (5.5)$$

so

$$R = r_o \qquad (5.6)$$

independently from temperature or any other parameters, the average distance between molecules interacting through a *Hooke* type force, is always the resting position of the spring r_o (!) What is changing then with temperature? Of course is the *spread* of this distance from the average r_o. Can you calculate the expression of this spread? I will give you a hint: you have to derive again, find the second derivative of the distribution and put it to zero. You will find two values that represent the range of variation of R around r_o and you will see that this does indeed depend on temperature!

5.2 Speed distribution in an ideal gas

In the previous chapters we derived the density distribution of the gas molecules for an ideal gas, in air. This time we consider about the speed of the molecules. To understand what is happening with the speeds, we have to remember these important things:

1. we are in *equilibrium*[26], so the velocity distribution must be the same everywhere in any point of the gas.
2. no forces are acting on the particles, only their own weight $-(mg)$
3. molecular *density* is known, is $n = n_0 e^{-(mgh)/kT}$

---

26) equilibrium: 平衡.

$$\left[-\frac{1}{kT}(\bar{k}R - \bar{k}r_o)\right] = 0 \tag{5.5}$$

となった．つまり

$$R = r_o \tag{5.6}$$

となる．

　この答えは，温度やほかのパラメータに依存していないということを示している．フックの法則でつながっている分子の平均の距離は，いつも自然長の r_o なのである！　では温度は何に作用するのだろうか．温度は平均距離 r_o からの分布の広がりに関係する．その広がりについて，あなたは計算できるだろうか？　ヒントをあげよう．分布をもう1回微分して，2階微分を $= 0$ にすればよい．そうすると，分布の広がり具合の R が r_o のまわりに2つ見つかるだろう．それが温度に依存している．

5.2　理想気体の速度分布

　前章では理想気体の密度分布を導いた．ここでは，分子の速度について考えてみよう．速度について理解する前に，重要な前提について確認しておこう．

1. 私たちは平衡状態の場合を考える．つまり，速度分布は気体のどこの場所でも同じである．
2. 分子自身の質量による力 $-(mg)$ の他は何も力を受けないとする．
3. 分子密度は既知で，$n = n_0 e^{-(mgh)/kT}$ である．

もしもこの3つの条件が成り立つ場合には，速度の分布が $e^{-(mg)h/kT}$ の項と何らかの関係があるということを直感的に理解できるだろう．では，図5.2に示すように2つの水平な線 $h = 0$ と $h = h$ を考えてみよう．$h = 0$ では，n_0 個の分子があり，$h = h$ では n 個の分子がいるとしよう．

5.2 Speed distribution in an ideal gas

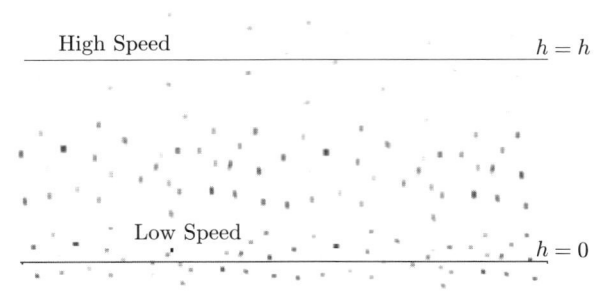

Fig. 5.2 Scheme[27] of the particles in motion in air. Less particles reach the higher level $h = h$ because of gravity. If we think in terms of speed, we can say that only the particles that have speed higher than u can reach this level. In other words, only those molecules moving at $h = 0$ with sufficient velocity can arrive at height h.

If this three conditions are true, it is intuitive to understand that the distribution of velocities should be somehow related to the term $e^{-(mg)h/kT}$. In fact, let's consider two horizontal lines $h = 0$ and $h = h$. As shown in Fig.5.2. At $h = 0$ we will have n_0 particles, instead at height $h = h$ there are n particles.

What's the ratio in number between the particles in these two levels? We know that higher up there are *less* particles (see condition "3"!) and that the ratio is: $e^{-(mgh)/kT}$. If there are no other forces than the weight $-(mg)$, then let's go down at level $h = 0$ and consider the velocities of the particles. For sure the particles missing at the level $h = h$ should be those that do not have enough kinetic energy to reach that place. What is this energy? It is the potential difference between the two heights, so:

$$E_k = dW = -mgh$$

If we set conventionally $u = 0$ as average velocity at the level $h = 0$, the kinetic energy to reach level h is $\frac{mu^2}{2} = mgh$.

So we can say that all those particles that do differ in kinetic energy of

---

27) scheme: 図. "schematic diagram: 概略図" のようによく使われる語.

5.2 理想気体の速度分布

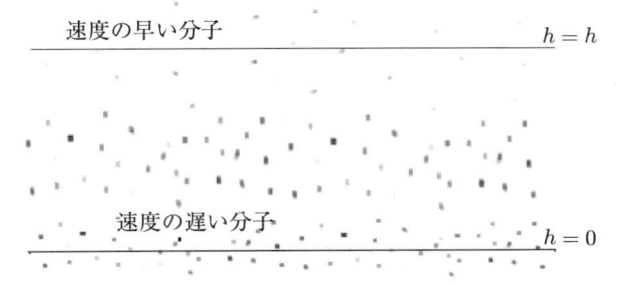

速度の早い分子　　　　　　　　　　　$h = h$

速度の遅い分子　　　　　　　　　　　$h = 0$

図 5.2　空気中での分子の動きの模式図．高い位置 $h = h$ に達する分子は重力の影響で少ない．速度について考えると，平均速度 u を超えるものだけがこの高いレベルに到達するとも言える．違う言葉でいうと，$h = 0$ から h まで上昇できるのに十分な速度をもった分子だけだとも言える．

この 2 つの水平線の間にいる分子数の比はどうなっているだろうか．4 章で学んだように，私たちは高いところに位置する気体分子の数は，低いところよりも少なくなるということを知っている．そして，その割合は $e^{-(mgh)/kT}$ である（条件 3）．もしも，分子の質量によって生じる $-(mg)$ という力以外の力が働いていない場合には，$h = 0$ での速度を考えればよい．$h = h$ で分子がなくなってしまうということは，その場所まで達するに十分な運動エネルギーがないということである．このエネルギーとは何だろう？　それは，2 つの高さの差から生じるポテンシャルによるものである．つまり，エネルギーは

$$E_k = dW = -mgh$$

と表すことができる．ここで，平均速度を u と呼ぶことにしよう．すると，h に達することができる運動エネルギーは $\frac{mu^2}{2} = mgh$ となる．

つまり，運動エネルギーが $mu^2/2$ と異なるこれらの粒子は高さ $h = h$ になったときになくなる．もしもこれが本当なら，$h = 0$ での粒子の運動エネルギーの割合を，私たちは次のように書くことができる．

$$\frac{n_{v>u}(h=0)}{n_{v>0}(h=0)} = e^{-(mgh)/kT} = e^{-mu^2/2kT} \tag{5.7}$$

ここで，$n_{v>u}$ は速度 u よりも速い速度で動いている粒子の数である．つまり，遅い速度の分子は高さ h に達するエネルギーが足りないということであ

$mu^2/2$ are those that will be missing at the higher level $h = h$. If this is true we can write the ratio of molecular density at $h = 0$:

$$\frac{n_{v>u}(h=0)}{n_{v>0}(h=0)} = e^{-(mgh)/kT} = e^{-mu^2/2kT} \qquad (5.7)$$

where $n_{v>u}$ means the number of particles with speed higher than u. In other words: molecules of lower speed have less energy to reach higher positions h. So, we simply obtained the *distribution of speed* within our gas...! To emphasize the distribution properties of the relation, in a number of books you can find the general *differential* form of this equation:

$$f(u)du = Ce^{-mu^2/2kT}du \qquad (5.8)$$

where we indicate with $f(u)$ the velocity ratio and with du the difference in average velocity. This is usually called the *Maxwell distribution* of velocities, because Maxwell derived it the first time in 1859. Notice the presence of the square on the independent variable u^2, this gives the distribution a shape of a Gaussian curve.

Remember: eq. (5.8) is the ratio of the number of particles of certain velocities in any single point of space. It is not the ratio of of velocities between $h = 0$ and $h = h$!

You may ask: "Wait a second! We know that higher up there are less particles because of the gravitational term $-mg$ in the Boltzmann distribution. So presumably higher up there is less speed too! Why you say that in any point of space the distribution is the same?!?". If you ask yourself that, you are perfectly right. That's why higher up in the mountain the air is thinner (n decreases) and temperature is colder (molecular speed is reduced). However, eq. (5.8) represent the only the *ratio* between velocities, not the absolute value. The average velocity of gas particles can change with height, still the ratio between velocities can be maintained for every point in space accordingly to eq. (5.8).

る．このように気体分子の速度分布は簡単に知ることができる．いくつかの本では，その分布の関係を強調するために，この式を微分した形で示すことがある．

$$f(u)du = Ce^{-mu^2/2kT}du \tag{5.8}$$

ここで $f(u)$ は速度の割合，du は平均速度からの差を示す．これは，通常マックスウェルの速度分布と呼ばれる分布である．なぜこう呼ばれているかというと，Maxwell が 1859 年に発見した法則だからである．ここで，独立変数として u^2 があることに着目しよう．これは，ガウス分布を導くものである．

式 (5.8) は，ある場所でのある一定の速度の分子の数の割合を示していることを覚えておこう．これは，$h = 0$ と $h = h$ の間にある速度の割合ではない．あなたはこう言うのではないだろうか．「ちょっと待って！ 上にいる分子が少ないのは，ボルツマン分布の重力 $-mg$ の項があるからではないか？ それ故に上にいる分子の速度は遅くなるのでは？ なぜ，どの場所においても分布が同じだと言ったのだろうか？」もしもあなたがそう質問するなら，あなたは全く以て正しい．まさにこれが，「山の上にあがると空気が薄くなり（密度 n の減少），温度が下がる（分子の速度が減る）こと」の理由である．しかし，式 (5.8) はただ速度分布の割合を示しているだけであって，絶対値ではない．平均速度は高さによって変わる．速度分布の割合は，どの場所にあっても式 (5.8) に一致する．

5.3 Brownian motion

Now we know that particles of a gas are moving around with different speeds. At any point in space we can meet a particle of lower or higher speed. Let's suppose we have a small ball immersed in a gas. This ball is a macroscopic object, so it is much bigger than the gas molecules. Let's call this particle a *pellet*. If we put the pellet in the middle of a gas, we know the gas particles move around and collide with the pellet continuously at any moment. This continuous collision is called also *bombardment*. So the pellet is bombarded from all directions by gas molecules. The molecules, do not have all the same speed or direction. However, they have a *distribution* of speed, as we learned above.

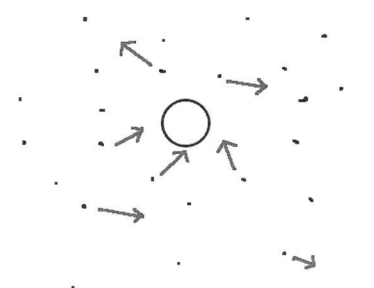

Fig. 5.3 A graphical example of chaotic *bombardment* of a *pellet* by gas particles.

Now, let's think about this: all these collisions come homogeneously from all directions, we may think that the total effect on the particle is zero. Is it? Well, it is not! Let's try to calculate this and understand why. Let's imagine that the pellet is set initially in the center of our coordinates defined by the point $R_0 = (0,0,0)$. Its position relative to this point is defined by the vector $\bar{R}$. After a single collision, the pallet moves to the left, or to the right, or up or any direction. Let's call this random movement $\bar{L}$. So the final position after $N = 1$ collision is

5.3 ブラウン運動

私たちは，気体の分子が，それぞれ異なる速さで動き回っていることを知っている．どの場所でも，速度の速い分子と，遅い分子を見つけることができる．それでは，気体中に小さなボールがある場合について考えてみよう．このボールは巨視的な物体である．もちろん，気体分子よりもずっとずっと大きい．このボールを，気体の真ん中に置くと，このボールの周りには気体分子が存在し，どの瞬間もボールに衝突している．この連続的な衝突は，衝撃とも呼ばれる．ボールは，あらゆる方向から気体分子の衝突を受けている．衝突する気体分子は同じ速度でも角度でもない．ただし気体分子の速度の分布については前に勉強した．

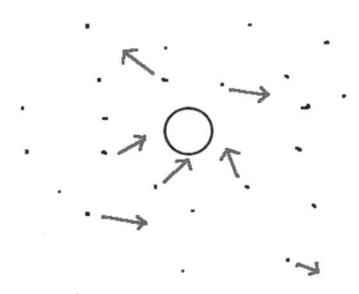

図 5.3 ボールに気体分子が無秩序に衝突する様子の模式図

今，この衝突はすべての方向から均一に起こっているとしよう．そうすると，あなたはボールに作用する力をすべて足し合わせるとゼロになると考えるだろうか．しかしそれは，正しくはない．なぜ正しくないのか，計算してみよう．まず，ボールを中心 $R_0 = (0,0,0)$ に置く．R_0 からの距離はベクトル $\bar{R}$ と定義する．1 回衝突した後，ボールは右，左，あるいは上などのどの方向へも動く可能性がある．このランダムな動きを $\bar{L}$ と定義しよう．つまり，$N = 1$ の後のボールの位置は

$$\bar{R}_1 = R_0 + \bar{L}$$

where R_0 is the start point. After two collision it will be

$$\bar{R}_2 = \bar{R}_1 + \bar{L}$$

and after three it will be

$$\bar{R}_3 = \bar{R}_2 + \bar{L}$$

Let's not forget that despite the fact that we always write $\bar{L}$, every time we have a different value. $\bar{L}$ is a displacement vector caused by a *random* force! In general we have

$$\bar{R}_N = \bar{R}_{N-1} + \bar{L}$$

What happens if we square right and left of this equation? We will have

$$\bar{R}_N \cdot \bar{R}_N = (\bar{R}_{N-1} + \bar{L}) \cdot (\bar{R}_{N-1} + \bar{L})$$

so we have

$$R_N^2 = R_{N-1}^2 + L^2 + 2\bar{R}_{N-1} \cdot \bar{L}$$

this equation is impossible to solve, because every collision $\bar{L}$ is different in value and different in directions. As usual in statistical mechanics, let's consider the *average* values of these terms.

$$< R_N^2 > = < R_{N-1}^2 > + < L^2 > + 2 < \bar{R}_{N-1} \cdot \bar{L} >$$

The first two terms are average of scalars (a *scalar* means a *number*, not a vector with direction and value, but only a value), so their average can be indicated for simplicity as only R_N^2, R_{N-1}^2 and L^2. Instead the remaining term $2 < \bar{R}_{N-1} \cdot \bar{L} >$ is the product of two vectors. We have to consider again that every time the collision occurs in different direction. These are random collisions, so they are spread homogeneously in all directions. The

5.3 ブラウン運動

$$\bar{R}_1 = R_0 + \bar{L}$$

となる．（R_0 はスタートの位置）．2回目の衝突後は

$$\bar{R}_2 = \bar{R}_1 + \bar{L}$$

3回目の衝突後は

$$\bar{R}_3 = \bar{R}_2 + \bar{L}$$

となるだろう．ここで，どれも $\bar{L}$ と書いているが，毎回違うということを忘れないようにしよう．$\bar{L}$ というのはランダムなのである．一般的に書くと，

$$\bar{R}_N = \bar{R}_{N-1} + \bar{L}$$

となる．

両辺を二乗すると

$$\bar{R}_N \cdot \bar{R}_N = (\bar{R}_{N-1} + \bar{L}) \cdot (\bar{R}_{N-1} + \bar{L})$$

となり，

$$R_N^2 = R_{N-1}^2 + L^2 + 2\bar{R}_{N-1} \cdot \bar{L}$$

となる．この方程式は解くことができない．なぜなら，すべての衝突 $\bar{L}$ は異なる値をもっていて，違う方向に進むからである．そこで，平均の値について考える．

$$<R_N^2> = <R_{N-1}^2> + <L^2> + 2<\bar{R}_{N-1} \cdot \bar{L}>$$

最初の2つの項はスカラーの平均である（スカラーというのは，ベクトルではなく値だけもっているということ）．そこで，これらの平均値は R_N^2，R_{N-1}^2 と L^2 によって示すことができる．残った $2<\bar{R}_{N-1} \cdot \bar{L}>$ は2つのベクトルの積である．私たちは再びここで，衝突は違う方向で起こるということについて考える．ただし，衝突はランダムに起こるので，すべての方向で均一に起こる．つまり平均のベクトル $\bar{L}$ はゼロになるべきである．この場合，

average vector $\bar{L}$ must be null. If this is the case, all the term $2 < \bar{R}_N \cdot \bar{L} >$ must be null. So at the end we have:

$$< R_N^2 > = < R_{N-1}^2 > + < L^2 > \qquad (5.9)$$

This equation is *recursive*, it means that the value of R at the Nth collision is equal to the value of R at the previous collision plus the average length of the collision L. If the collision happens 10 times, we have $R = 10 * L$. So, finally, we can write

$$R^2 = N < L^2 > \qquad (5.10)$$

that means that the distance of the pellet from the start point $R_0 = (0,0,0)$ increases like the following:

$$R = \sqrt{N} < L > \qquad (5.11)$$

If we can estimate the average length of a free movement $< L >$, we will know the total distance after N collisions. This is called *Brownian drift*. The thing you must notice is that even if the collisions are totally random, this value is not zero!

We learned from eq. (5.11) that the distance R increases with the number of collisions. But -of course- in real life, we do not know the number of collisions...! In a *real* problem, what shall we do? What is the real variable that we can measure, instead of the number of collisions N? Clearly, on average, we have a constant number of collisions per unity of time. So, N is proportional to time: $N = rt$, where r is the *rate of collision* (number of collisions per unity of time).

The distance from the start point increases with time. From eq (5.10) we can state:

$$< R^2 > = N < L^2 > = rt < L^2 > = \alpha t \qquad (5.12)$$

5.3 ブラウン運動

すべての項 $2<\bar{R}_N \cdot \bar{L}>$ はゼロになるべきであり，つまり，最終的に，

$$<R_N^2>=<R_{N-1}^2>+<L^2> \tag{5.9}$$

となる．

この式は帰納的である．なぜなら N 回目の衝突での R と，その前の衝突の R の値に L の平均長さを足したものが同じだからである．もしも 10 回衝突が起こったとしたら，$R=10 \times L$ となる．つまり，最終的に

$$R^2 = N<L^2> \tag{5.10}$$

と書くことができる．これは，ボールが原点 $R_0 = (0,0,0)$ からの距離を下の式のように増やしているということを意味している．

$$R = \sqrt{N}<L> \tag{5.11}$$

つまり，私たちがもし平均移動距離 $<L>$ を見積もることができれば，N 回の衝突後での距離を知ることができる．これは，**ブラウン・ドリフト**と呼ばれている．衝突はまったくランダムに起こるが，この値はゼロにならないということに着目すべきである．

私たちは式 (5.11) から，距離 R が衝突の回数とともに増えることを学んだ．しかしもちろん，現実の世界では，その衝突の回数を知ることはできない．現実の問題として取り組むとき，私たちはどうしたらよいのだろうか．衝突回数 N の代わりに私たちが実際に測定することのできる変数は何だろうか．もちろん，平均をとれば，単位時間の中の衝突回数を知ることはできる．つまり，N は $N = rt$ のように時間に依存している．ここで r は単位時間内に衝突する回数，**衝突頻度**を示している．

原点からの距離も時間とともに増加する．式 (5.10) より，

$$<R^2>=N<L^2>=rt<L^2>=\alpha t \tag{5.12}$$

を得ることでできる．ここで α は $r<L^2>$ である．前に書いたように，$\bar{L}$ は

Where α is then $r < L^2 >$. As said above, the $\bar{L}$ is the average movement that the particle does between collision. We can call this *mean free path*[28] and indicate it with the letter λ. So $\lambda = <\bar{L}>$, it represent the average length of a *free* movement the particle does between collisions. The term *free* is given because we suppose that during this movement the particle is totally free of forces, so the movement is a straight line of uniform speed.

Now, let's try to estimate this mean free path. Well, if we know the average velocity of the particle (let's call it v_o), then we can easily say:

$$\lambda = \frac{v_o}{r} \qquad (5.13)$$

or $\lambda = v_o \tau$ if $\tau = \frac{1}{r}$ (we can call τ *mean free time*[29]). Please notice that the velocity v_o is not the velocity of the gas molecules, but the velocity of our *pellet* macro-sized particle that is bombarded by the molecules.

If you remember the laws of ideal gas, you know that the average velocity of the gas molecules is known, it depends on the temperature and using eq. (1.11):

$$\frac{1}{2}mv^2 = \frac{3}{2}kT \qquad (5.14)$$

Gas molecules of this average velocity will collide with the particle. For the conservation of momentum[30], the particle -on average- will move away with this speed:

$$m_o v_o = mv \qquad (5.15)$$

where m_o is the mass of the pellet, then taking in account eq. 5.14 we can write:

---

28) mean free path: 平均自由行程. 分子が他の分子と衝突してから，次の分子に衝突するまでに進む平均距離.
29) mean free time: 平均自由時間. 分子が他の分子と衝突してから次の衝突までに要する平均時間のこと．
30) momentum: 運動量.

衝突後の粒子の平均の動きを示している．私たちはこれを平均自由行程と呼ぶ．そして，λという文字で表す．つまり，$\lambda = <\bar{L}>$となる．これは衝突前後での自由な動きの平均移動距離である．この自由というのは，粒子が全体的にみて，力から自由であるということに由来する．つまり，一定の速度で一直線に進む．

さて，平均自由行程を見積もってみよう．もしも分子の平均速度v_oを知っているとすると，

$$\lambda = \frac{v_o}{r} \tag{5.13}$$

または$\tau = \frac{1}{r}$の場合$\lambda = v_o \tau$となる．τは，平均自由時間と呼ぶ．ここで，v_oが気体の分子の速度ではなく，分子によって衝突されているボールのような巨視的なサイズのものの速度であることに注意したい．

もしもあなたが，理想気体の状態方程式を覚えていたとすると，平均の気体速度はわかるし，それが温度に依存していることも知っている．式 (1.11) を使うと，

$$\frac{1}{2}mv^2 = \frac{3}{2}kT \tag{5.14}$$

となる．この平均速度をもった気体分子がボールと衝突を起こす．運動量保存則によって，ボールは（平均では）この速度で離れていく．

$$m_o v_o = mv \tag{5.15}$$

ここでm_oはボールの質量である．そうすると，式 (5.14) により，私たちは

$$v_o = \frac{m}{m_o}v$$
$$\frac{3}{2}kT = \frac{1}{2}m\left(\frac{m_o}{m}v_o\right)^2 \tag{5.16}$$

と書くことができる．

また適切に簡略化することにより，粒子の平均速度の式を次のようにすることができる．

$$v_o = \frac{m}{m_o}v$$
$$\frac{3}{2}kT = \frac{1}{2}m\left(\frac{m_o}{m}v_o\right)^2 \tag{5.16}$$

doing the proper simplifications, you can obtain easily the expression for the average particle velocity:

$$v_o = \frac{\sqrt{3kTm}}{m_o} \tag{5.17}$$

If we substitute this in our previous expression for λ eq. 5.13, we have finally

$$\lambda = \frac{\sqrt{3kTm}}{m_o r} \tag{5.18}$$

Now substituting this in the expression eq. (5.12) that we where looking for, we have

$$R^2 = \frac{3kTm}{m_o^2 r}t \tag{5.19}$$

What is the rate of collision r? For sure it is something related to the temperature of the gas. If the gas has more energy, presumably there will be more collisions. However, this parameter must be connected also to the geometrical dimensions of the particle. If the particle is bigger will receive many collision per second, if it is very small just a few. Let's suppose that the gas molecules are so small that their size can be neglected compared to the dimension of the particle. Then the probability of having a collision will depend only on the size of the particle and the molecular density per unit of volume n. If the section of the particle has an area σ (usually called *cross section*[31]) then the variation of the number of collisions ΔN will be

$$\Delta N = \sigma n \Delta x \tag{5.20}$$

where Δx is an infinitesimal displacement in space.

---

31) cross section: 断面積. 分子が衝突を起こす確率のこと.

5.3 ブラウン運動

$$v_o = \frac{\sqrt{3kTm}}{m_o} \tag{5.17}$$

ここでもし，λ に関して前の式 (5.13) を使うと仮定すると，最終的な式

$$\lambda = \frac{\sqrt{3kTm}}{m_o r} \tag{5.18}$$

を得ることができる．ここで，式 (5.18) を式 (5.12) に代入すると，

$$R^2 = \frac{3kTm}{m_o^2 r} t \tag{5.19}$$

となる．

衝突頻度の r とはなんだろうか．何か気体の温度と関係しているというのは確かである．もし気体分子のエネルギーがあがると，衝突の頻度は増える．しかし，このパラーメータはボールの幾何学的な次元にも依存する．つまり，粒子が大きければ単位時間あたりの衝突回数は増えるし，小さければ減る．では，気体分子がとても小さく，そのサイズが無視できるという条件で考えてみよう．そうすると，衝突頻度はボールのサイズと，単位体積あたりの気体分子の密度 n に依存する．もしも粒子の断面の面積が σ だとすると，（これを通常，**断面積**と呼ぶ），Δx という空間の微小な変位で起こる衝突の回数の変化 ΔN は次式のようになる．

$$\Delta N = \sigma n \Delta x \tag{5.20}$$

今，x 方向に，気体分子が $\frac{1}{2}kT$ という運動エネルギーをもっていたとする．これは $\frac{1}{2}mv^2$ と同じである．また，無限小変位 dx の方向が分子の動く向きと同じであると仮定する．すると $\Delta N = dN = \sigma n v dt$ となり，

$$r = \frac{dN}{dt} = \sigma n v \tag{5.21}$$

となる．すると $\frac{1}{2}kT = \frac{1}{2}mv^2$ より簡単に，

$$r = \frac{dN}{dt} = \sigma n \sqrt{\frac{kT}{m}} \tag{5.22}$$

を得る．すると

Now, in the x direction, the gas has a kinetic energy $\frac{1}{2}kT$ that is equal to $\frac{1}{2}mv^2$. We can assume that the infinitesimal displacement dx correspond to the molecular movement in the same direction. So, $\Delta N = dN = \sigma n v dt$. The rate is the number of collisions per unity of time, then

$$r = \frac{dN}{dt} = \sigma n v \qquad (5.21)$$

Because $\frac{1}{2}kT = \frac{1}{2}mv^2$ easily we obtain:

$$r = \frac{dN}{dt} = \sigma n \sqrt{\frac{kT}{m}} \qquad (5.22)$$

then:

$$R^2 = \frac{3\sqrt{kT}m^{3/2}}{m_o^2 \sigma n} t \qquad (5.23)$$

This equation shows how the Brownian motion depends on temperature and other important parameters. The group of terms $\frac{3\sqrt{kT}m^{3/2}}{m_o^2 \sigma n}$ represents our parameter α that was defined previously, eq (5.12). It simply tells us how fast is the Brownian motion; it is a kind of Brownian "speed". It is not a real speed because R is the distance from a center point $R_0 = (0,0,0)$, so α tells us how far are we from that point, but does not tell us exactly where we are! We only know that the pellet goes away in a random direction within a radius R given by eq (5.23). Try to imagine for example that the particle m_o becomes very heavy, or its size gets bigger (σ increases)... the total movement becomes slow because of the inertia or because the total amount of collisions average out to zero. The same will happen if the molecular density increases. If instead the temperature T grows, or the gas molecules are heavier, the Brownian movements get faster and quickly the particles run away, as it is logical to be. Our simple model makes sense!

$$R^2 = \frac{3\sqrt{kT}m^{3/2}}{m_o^2 \sigma n} t \tag{5.23}$$

が得られる．この式はブラウン運動が温度やほかの重要なパラメータにどのように依存しているかを示している．$\frac{3\sqrt{kT}m^{3/2}}{m_o^2 \sigma n}$ の項は以前に式 (5.12) で定義したパラメータ α を表しており，α はブラウン運動がどのくらい速いかを簡潔に示してくれる．これはブラウン速さといってもいい．ただしこれは真の意味での速さではない．なぜならば R は原点 $R_0 = (0,0,0)$ からの距離だからである．つまり，α はこの点からどのくらい遠くにいるかを教えてくれるが，正確な場所を私たちに教えてくれない．ランダムに進んだボールの半径 R は式 (5.23) によって与えられる半径 R の範囲内にある．粒子 m_o がとても重い場合を想像してみよう．あるいは，サイズがとても大きく断面積 σ が増加した場合について考えてみよう．移動は慣性，あるいは衝突が減ることによりにより全体的にゆっくりになる．同じことが分子密度が増えても起こる．もしも温度 T が増加したり，または気体分子の質量 m が重くなると，ブラウン運動は早くなり，粒子は速く動きまわる．これは論理的にも筋がとおっている．私たちの簡単なモデルが有意義だったことを示している．

Example 5.2

Let's consider a particle of mass $m_o = 10^{-20}$ kg that is inserted in Argon gas at room temperature $T = 300$ K. We suppose that the molecule has a cross section of 10^{-18} square meter and that argon molecular density is $n = 10^{24}$ molecules per cube meter. After a duration of $t = 100$ seconds, can you calculate the average distance that this particle run from the starting point $R = (0,0)$ due to Brownian motion?

The parameter we use are the ones indicated here:

$$m_o = 10^{-20} \text{ kg}$$
$$T = 300 \text{ K}$$
$$n = 10^{24} \text{ m}^{-3}$$
$$\sigma = 10^{-18} \text{ m}^2$$
$$m = \frac{18 \times 10^{-3}}{6 \times 10^{23}}$$

then we apply equation (5.23):

$$R^2 = \frac{3\sqrt{kT}m^{\frac{3}{2}}}{m_o^2 \sigma n} t \qquad (5.24)$$

Using $t = 100$ seconds, we obtain:

$$R^2 = \frac{3\sqrt{1.38 \times 10^{-23} \times 300} \times (3 \times 10^{-26})^{\frac{3}{2}}}{(10^{-20})^2 \times 10^{-18} \times 10^{24}} \times 100 \qquad (5.25)$$

which is:

$R = 1$ micrometer

5.4 Thermal noise

Brownian motion is caused by the temperature. We know that the motion of the particles is energy, and that the temperature we feel in an object is simply the kinetic energy of the molecules of that object. For this reason the

例題 5.2

質量 $m_o = 10^{-20}$ kg の分子を室温 ($T = 300$ K) のアルゴンガスの中に入れた場合を考えよう．この粒子は断面積 10^{-18} m^2，アルゴンの密度は $n = 10^{24}$ m^{-3} である．100 s の後，この物質がスタートの位置 $R = (0,0)$ から平均でどのくらい動いたかを考えよう．

パラメータは下記のとおりである．

$$m_o = 10^{-20} \text{ kg}$$
$$T = 300 \text{ K}$$
$$n = 10^{24} \text{ m}^{-3}$$
$$\sigma = 10^{-18} \text{ m}^2$$
$$m = \frac{18 \times 10^{-3}}{6 \times 10^{23}}$$

すると式 (5.23)

$$R^2 = \frac{3\sqrt{kT}m^{3/2}}{m_o^2 \sigma n} t \tag{5.24}$$

に，このパラメータと $t = 100$ s を代入すると

$$R^2 = \frac{3\sqrt{1.38 \times 10^{-23} \times 300} \times (3 \times 10^{-26})^{\frac{3}{2}}}{(10^{-20})^2 \times 10^{-18} \times 10^{24}} \times 100 \tag{5.25}$$

これを解くと

$$R = 1 \text{ } \mu\text{m}$$

となる．

5.4 熱雑音

ブラウン運動は温度によって引き起こされる．私たちは，粒子の動きというものがエネルギーであるということを知っている．また，私たちが物質に感じる温度というものは，単純に物体の分子の運動エネルギーであるということを

5.4 Thermal noise

Brownian motions have consequences generally, on all possible systems! For examples lets consider again our pellet immersed in a gas. We know that the kinetic energy of the gas, for simplicity we consider only one dimension x, is

$$E_{gas} = \frac{1}{2}kT$$

we know that this must equal the average kinetic energy of the pellet, so

$$\frac{1}{2}m<v_x^2> = \frac{1}{2}kT$$

where m is the mass of the pellet. So we can say that

$$<v_x> = \pm\sqrt{\frac{kT}{m}}$$

What does this means in practice? It means that because of the thermal energy, whatever is the velocity of the pellet, there is always an average thermal velocity that adds up to the total velocity of the pellet. This velocity is very small, but it always exists. For a pellet of one gram it is (let's use MKS):

$$m = 0.001 \text{ kg} \tag{5.26}$$
$$k = 1.38 \times 10^{-23} \text{ J/K}$$
$$T = 300 \text{ K}$$
$$<v_x> = \pm 2 \times 10^{-9} \text{ m/s}$$

very small, but not zero. This fact also has implications when we want to measure the velocity of the pellet in the gas, this velocity can be big, because there will be other external forces we put on the pellet. However, our measurement of the pellet's speed will be disturbed. We cannot know the speed of the pellet, with a precision better than $<v_x>$. In mathematical terms: if the real velocity is V_0, what we actually measure is

知っている．これにより，ブラウン運動は一般的にすべての系で成り立つ．例えば，気体の中にボールを置いてみよう．私たちは気体分子の運動エネルギーを知っている．簡単にするために一次元の x 方向だけ考えると，

$$E_{gas} = \frac{1}{2}kT$$

となる．このエネルギーはボールの平均の運動エネルギーなので，

$$\frac{1}{2}m<v_x^2> = \frac{1}{2}kT$$

となる．ここで m とはボールの質量である．つまり，

$$<v_x> = \pm\sqrt{\frac{kT}{m}}$$

となる．これはボールにとって，どんな意味があるのだろうか．この式が意味することは，ボールの速度がどうあれ，熱エネルギーによって，平均的にみて熱的な速度が加わり，その速度は速くなるということである．この速度の変化分は微小だが，しかし，常に存在している．1gのボールでは，

$$m = 0.001 \text{ kg} \tag{5.26}$$
$$k = 1.38 \times 10^{-23} \text{ J/K}$$
$$T = 300 \text{ K}$$
$$<v_x> = \pm 2 \times 10^{-9} \text{ m/s}$$

となる（MKS単位系を使おう）．とても小さいけれども，ゼロではない．この事実は，気体のなかに置いたボールの速度を私たちが測るとき，その速度は外部からの力によって大きくなっている，ということを意味している．私たちのボールの速度の測定は，この力によって妨げられる．つまり私たちは，ボールの速度について，$<v_x>$ よりも正確に測定することはできないのである．数学的にいうと，もしも真の速度が V_0 だとすると，私たちが測るのは

$$V = V_0 \pm <v_x>$$

となる．この効果こそが**熱雑音**と呼ばれるものである．この事実は，どんな系

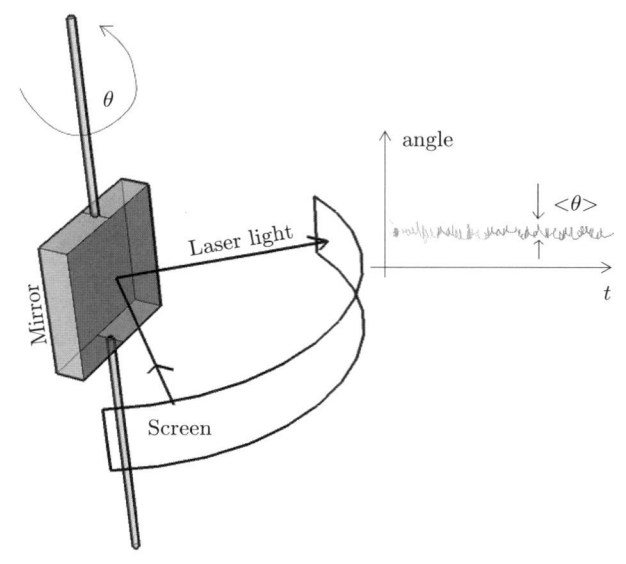

Fig. 5.4 The scheme of a light-beam galvanometer, a Laser light is reflected by a mirror suspended on a very low friction string. The reflection at angle θ is affected by an error of $<\theta>$ due to thermal noise, as represented in the graph.

$$V = V_0 \pm <v_x>$$

This is the effect of the *thermal noise* on our measure.

This fact applies in any system we can think of. Where there is temperature (thermal energy) there is this thermal noise that affect measurements.

For example we can consider a mirror suspended by a small string, like in figure 5.4. In this case the thermal energy $\frac{1}{2}kT$ will be equal to the rotational energy of the mirror. If you remember from the physics classes, this energy is $\frac{1}{2}I\omega_0^2 <\theta^2>$, ($I$ is the moment of inertia, ω_0 is the angular frequency and θ is the displacement angle), these two should be equal so we have:

$$<\theta> = \pm \frac{\sqrt{kT/I}}{\omega_0}$$

This effect of thermal noise can be verified experimentally in the laboratory!

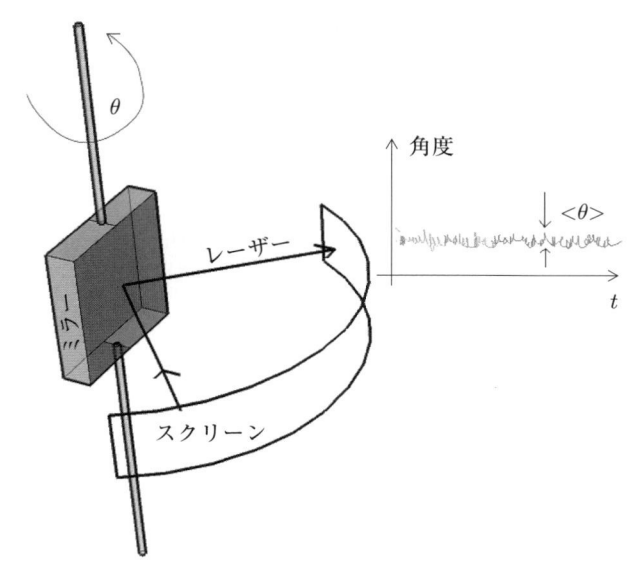

図 5.4 光を利用した検流計の模式図. レーザー光はとても摩擦の小さい糸につり下げられた鏡によって反射する. θ の角度での反射は $<\theta>$ の誤差によって影響される. それは熱雑音によってひき起こされ, グラフに現れる.

でも成り立つ. 温度がゼロでなければ（つまり熱エネルギーがあれば），そこには熱雑音が測定に入ってくるということである.

例えば, 図 5.4 のように糸に鏡をつり下げた場合について考えてみよう. このとき, 熱エネルギー $\frac{1}{2}kT$ は鏡の回転エネルギーと等しくなる. もしも物理の授業でならったことを覚えていると, このエネルギーは $\frac{1}{2}I\omega_0^2 <\theta^2>$ である. I は慣性モーメント, ω_0 は角振動数, θ は変位角である. この 2 つは等価なので,

$$<\theta> = \pm \frac{\sqrt{kT/I}}{\omega_0}$$

という式を導くことができる. これは, 熱雑音を実験室で実験的に証明できるということを示している. いろんな例を出してきたが, 最後に電子回路について話すと, コイルに蓄積されたエネルギー $\frac{1}{2}LI^2$ は, 同じ理由で $\frac{1}{2}kT$ と等しくなる. これは $<I> = \sqrt{\frac{kT}{L}}$ のときに

A variety of examples are possible, for example let's consider electric circuits. The energy accumulated in a inductance $\frac{1}{2}LI^2$ will be equal to $\frac{1}{2}kT$ with similar consequences

$$I = I_o \pm <I>$$

where $<I> = \sqrt{\frac{kT}{L}}$. Can you think to some other examples in another fields?

5.5 Evaporation

As you remember from previous chapters, equation (4.20) expresses the fact that the probability of finding a particle at a certain distance is related to the potential energy $(P.E.)$.

$$f = const\, e^{-P.E./kT}$$

This consideration is usually called *kinetic theory* because, as you remember, equation above is derived from the fact that we demonstrated that the kinetic energy in one dimension (for example the dimension "x") $E_k = \frac{1}{2}m<v_x^2>$ is equal to the thermal energy

$$\frac{1}{2}m<v_x^2> = \frac{1}{2}kT \qquad (5.27)$$

Now, let's consider the evaporation of a liquid, and let's try to apply the same concepts of kinetic theory, in order to find the equation of such system. Let's consider a closed box filled with liquid at a certain temperature. Because there is temperature, we know from kinetic theory that there will be molecular speed. If so, why the molecules do not fly away like a gas? Well, because there is a force keeping them together. We do not know exactly the mathematical formula for this force, but let's assume that we know the shape of it. The force is represented in figure 5.5.

From the shape of the curve you can understand that if the temperature is such that $E_k(=\frac{1}{2}kT) < \phi$, then the particles seem to be trapped near

$$I = I_o \pm <I>$$

となるのと同様である．あなたは，他の分野で，同様の例を思いつくだろうか？

5.5 蒸発

もし皆さんが前の章を覚えていたとしたら，ある場所で粒子を見いだす確率を表している式 (4.20) がポテンシャルエネルギー ($P.E.$) に依存しているということを知っているだろう．

$$f = const\, e^{-P.E./kT}$$

この考察は通常**気体分子運動論**と呼ばれる．なぜなら，（覚えていると思うけれども），上の式は1次元（たとえば x 方向など）の運動エネルギーが熱エネルギー

$$\frac{1}{2}m<v_x^2> = \frac{1}{2}kT \tag{5.27}$$

と等しいということから導いたものだからである．

それではここで，液体の蒸発について考えてみよう．そして，気体分子運動論を応用して，蒸発をどのように考えたらいいか，取り組んでみよう．まず，ある一定の温度の液体が四角い閉じた箱に満たされている状況を考えよう．温度がゼロではないので，気体分子運動論から分子の速度はゼロではないことが分かる．もしそうならば，なぜ分子は気体分子のように自由に飛んでいかないのだろうか？　それは分子には互いに結びつけている力があるからである．この力について，具体的な数式は今は知らないとしても，その形を推定することはできるのではないだろうか．力は図 5.5 のように表されるはずである．

この曲線の形から，もしも温度が $E_k(=\frac{1}{2}kT) < \phi$ の状態にあるならば，分子は ϕ よりも小さいエネルギーしか持っていないので，分子が互いに近い距

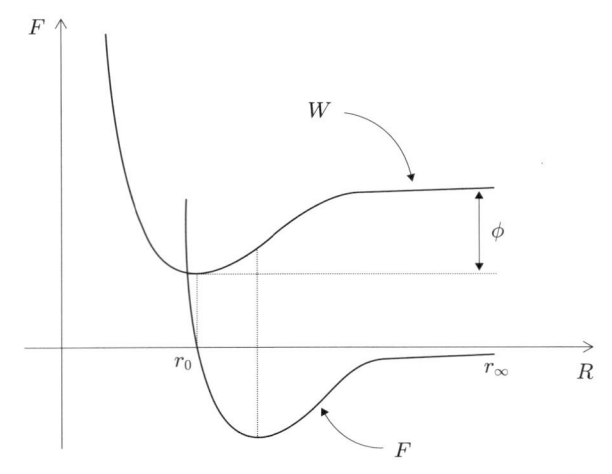

Fig. 5.5 The potential energy between liquid molecule (upper curve). We remember that force is $F = -\frac{dW}{dx}$ and it is represented by the lower curve in the picture. ϕ represents the potential energy necessary to remove the particle from the attraction. This energy is often called *work function*[32].

each other. So they cannot escape in the gas state.

Is this really right? Well, no! We know that the distribution of velocities is an exponential decreasing curve that never reaches zero (equation 4.19). So, some particle will have a speed higher than the average $E_k = \frac{1}{2}kT$. So -for sure- these few molecules will leave the liquid and become gas. Thus, in our box we will have a liquid *in equilibrium* with a certain small quantity of gas. This gas is also called *vapour*. The energy ϕ necessary to leave the liquid and become gas, is called *work function*.

Now, let's consider this problem: in our box of volume V, we have a vapour with n molecules per unity of volume and a certain amount of molecules in liquid phase. How many molecules will be in the vapour phase, compared with the number that are in liquid?

---

32) work function: 仕事関数.

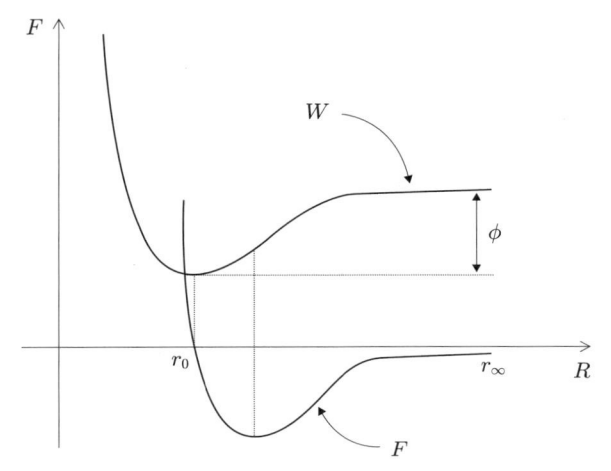

図 5.5 液体の分子間に働くポテンシャルエネルギー（上の曲線）と力 $F = -\frac{dW}{dx}$（下の曲線）．ϕ は r_0 と r_∞ でのポテンシャルエネルギーの差で，分子同士を引力から引き離すのに必要なエネルギーを示している．このエネルギーは，通常，仕事関数と呼ばれる．

離で存在するということが理解できるだろう．つまり，気体になって自由に飛び回ることができないということを示している．

　これは正しいのだろうか？　いや，正しくない．なぜなら，私たちは速度の分布が式 (4.19) によると，決してゼロとならないという結果を知っているからである．つまり，いくつかの分子は平均の $E_k = \frac{1}{2}kT$ よりも高いエネルギーを持っているということである．つまり，当たり前だが，いくつかの分子は液体から飛び去って気体分子になる．だから，もしも箱に液体が入っていると，いくらかの量の気体分子と平衡状態にあるわけである．この気体分子は蒸気と呼ばれている．液体から気体になるには，ϕ より大きいエネルギーを必要とする．これは**仕事関数**と呼ばれる．

　さて，次の問題について考えてみよう．蒸気相と液相の分子が存在する体積 V の箱を考えよう．ここで蒸気相では単位体積あたり n 個で分子が存在しているとする．このとき，液体に含まれる分子の数と比べて，蒸気相にある気体分子の数はいくつだろうか．

5.5 Evaporation

The solution is not so difficult. Lets suppose that every molecule has a volume V_a. Then the number of molecules per unity of volume in the liquid is $1/V_a$. So we have $n_1 = n$ molecules per unity of volume in vapour phase, and $n_2 = 1/V_a$ molecules per unity of volume in liquid phase.

Now we have to remember equation (4.19). Lets consider the two situations: gas phase and liquid phase. We write eq. (4.19) for each case:

$$n_1 = n_0 \, e^{-W_1/kT} \tag{5.28}$$

$$n_2 = n_0 \, e^{-W_2/kT} \tag{5.29}$$

now we simply substitute and have

$$n = n_0 \, e^{-W_1/kT} \tag{5.30}$$

$$\frac{1}{V_a} = n_0 \, e^{-W_2/kT} \tag{5.31}$$

n_0 is not relevant.

It is simply the density of molecules for the initial potential. It is a value that we do not know, and we do not care to know. In fact, lets divide eq. (5.30) by eq. (5.31) and we have

$$nV_a = e^{-(W_1-W_2)/kT} \tag{5.32}$$

the difference in potential energy $W_1 - W_2$, is the work function, also called ϕ, so we have finally:

$$nV_a = e^{-\phi/kT} \tag{5.33}$$

What does it mean? It means that the ratio of the density in liquid phase and in gas phase, is proportional to the exponent of some energy (the work function) divided by kT. In other words, if we have a big work function ϕ, this exponent is a big number, and small variations of temperature, make big variations of this ratio.

5.5 蒸発

この答えはそれほど難しくない．まず，分子がそれぞれ体積 V_a を持っているとする．すると，液体の単位体積あたりの分子の数は $1/V_a$ 個である．そこで，蒸気相に $n_1 = n$ の分子が単位体積あたり存在し，液体の単位体積あたり $n_2 = 1/V_a$ の分子が存在するとしよう．

つぎに，式 (4.19) を思い出そう．気相と液相，2 種類の状況を考える．式 (4.19) をそれぞれの場合について，

$$n_1 = n_0\, e^{-W_1/kT} \tag{5.28}$$

$$n_2 = n_0\, e^{-W_2/kT} \tag{5.29}$$

と書こう．ここで n_1, n_2 に置き換えると，

$$n = n_0\, e^{-W_1/kT} \tag{5.30}$$

$$\frac{1}{V_a} = n_0\, e^{-W_2/kT} \tag{5.31}$$

となる．n_0 は重要な値ではない．単に最初のポテンシャルでの分子の密度を示したもので，その値を知らなくても大丈夫である．では，式 (5.30) を式 (5.31) で割り算しよう．すると

$$nV_a = e^{-(W_1-W_2)/kT} \tag{5.32}$$

を得る．ポテンシャルエネルギーの差 $W_1 - W_2$ とは，つまり仕事関数のことである．従って，ϕ と書くことも可能で，

$$nV_a = e^{-\phi/kT} \tag{5.33}$$

となる．これは一体何を意味するのだろうか．液体と気体の密度の比は何かのエネルギー（仕事関数）を kT で割った値の指数関数に比例しているということである．言い換えると，もしも仕事関数 ϕ が大きかったら，指数は大きな値となり，温度の小さな変化でもこの比を大きく変化させる．

Anyways, please remember that this equation is the result of our model. So there are *assumptions* we made, we must remember that:

1. we are in *equilibrium* so the velocity distribution must be the same everywhere.
2. the volume occupied by one particle is not constant! If the temperature changes, also the volume occupied by a particle changes (the liquid *expands*).
3. The real situation is much more complicated, and we cannot obtain a formula. We use this equation (5.33) because it is simple and it is demonstrated that is valid in good approximation.

Lets remember a most important concept in physics. Everything we know are results of our models. Our models are not the true reality of physics, but an approximation of it.

Now let's try a very important test. Let's use another model, and describe again evaporation. We have our liquid in the closed box. Some of the molecules have escaped from the liquid and become vapour (gas). Of course, some of these gas molecules will hit the surface of the liquid, and become liquid again. How many are them?

Let's think small and imagine a small volume of gas, of area dV. This small volume of gas it is located exactly adjacent to the surface of the liquid. How many molecules per unity of time, will hit the liquid in this small volume? Well, we know that the gas density per unity of volume is n. Then we know that the total number of molecules N in this volume is -of course- $N = ndV$. If this volume has an area A, then dV can be written also as Adx, where dx is the movement toward the liquid in a short time dt. We know that the average speed of the molecules is v, so we can write

5.5 蒸発

いずれにせよ，この式が私たちのモデルの結果であることを覚えていて欲しい．そこで，私たちは下記の仮定について覚えておく必要がある．

1. 平衡状態を考えているので，速度分布はどの場所でも等しくなる．
2. 1つの分子が占有する体積は <u>一定ではない</u>！ もしも温度が変わると，分子によって占有される体積も変化する．（液体は膨張する）
3. 現実の状況はより複雑で，式で表すことは難しい．私たちが式 (5.33) を使ったのは，簡単なのによい近似を示すからである．

では，物理学にとって，最も大切な概念を思い出そう．それは導いた結果は私たちのモデルを反映したものだ，ということである．私たちが考えたモデルというのは，現実の世界をすべて反映しているわけではなく，ひとつの近似である．

それでは，異なるモデルを考えて，もう一度蒸発について記述してみるという重要なテストにチャレンジしてみよう．まず，箱に液体を準備しよう．液体の分子のいくつかは蒸発して，蒸気になっているとする．もちろん，気体分子のいくつかは，液体の表面にぶつかり，そして液体にもどるものもある．いくつの分子がそうなるのだろうか？

ある小さな空間 dV をまず考えてみよう．この空間は，液体表面のごく近くにあるとする．この微小空間では単位時間あたりに，いくつの分子が液体の表面にぶつかるだろうか．単位体積あたりの気体の密度は n なので，微小空間にいる分子の総数 N はもちろん $N = ndV$ となる．この空間は面積 A を持っているので，dV というのは Adx とも書ける．ここで dx は微小時間 dt の間に液体への移動である．そうすると，分子の平均の速度 v というのは

$$N = nAdx$$
$$N = nAvdt \tag{5.34}$$

となる．もしも単位時間を 1 s と決定し，単位面積を 1 m^2 とすると，液体になる分子は

5.5 Evaporation

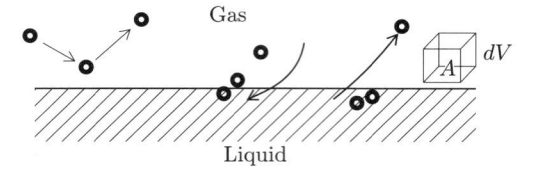

Fig. 5.6 A scheme of the condensation-evaporation process. A liquid is in a box a temperature T. Some molecules of the liquid win enough energy to escape and become a gas. Some gas molecules, on the contrary, hit the surface and condense in the liquid.

$$N = nAdx$$
$$N = nAvdt \tag{5.34}$$

If we choose a unity of time (one sec) and a unit of area (one square meter) then the number of molecules that condensate on the liquid per second per square meter is:

$$N_c = nv \tag{5.35}$$

Condensation, means a transformation of phase, from gas to liquid. What about the opposite transformation *evaporation* from liquid to gas? Our question is: how many molecules per unit of time leave the liquid?

If a liquid molecule acquire enough energy, it can win the energy gap dW and become a gas molecule. We know from previous kinetic theory (Boltzmann Law, eq. 4.19) that the probability that this happens is $e^{-dW/kT}$. Let's do some simple considerations: we understand that the number of molecules per unit of time leaving the liquid (and becoming a gas) should be proportional to the number of atoms near the surface, per unit of area, divided by the time it takes to leave the surface, multiplied by the probability to leave the surface. In mathematical terms this is:

$$N_e = \left(\frac{N}{A}\right)\left(\frac{1}{t_e}\right) e^{-dW/kT} \tag{5.36}$$

In the case of liquid, every molecule is packed near each other, the number of atoms per unit of area is simply A/A_s, where A is the total area and A_s

5.5 蒸発

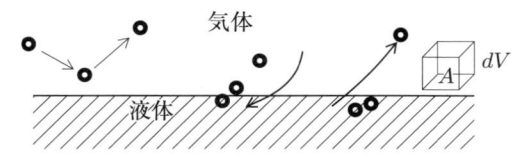

図 5.6 凝縮と蒸発のプロセス．液体は温度 T の箱に入っている．いくつかの分子は液相から脱出するのに十分なエネルギーをもっていて，気体となっている．いくつかの気体分子は，逆に表面にぶつかり，凝縮して液体となる．

$$N_c = nv \tag{5.35}$$

となる．

凝縮というのは，気相から液相に相転移が起こるということを意味している．逆の場合（液相から気相）は蒸発という．私たちの問題は，単位時間にいくつの分子が液体から飛び去るかということである．

もしも液相の分子が十分なエネルギーを得ると，エネルギーギャップ dW を乗り越えて気相分子になる．すでに運動エネルギーについて知っているので（ボルツマンの法則，式 (4.19) 参照のこと），この蒸発が起こる確率は $e^{-dW/kT}$ であるとわかる．簡単な考察をしてみよう．液面から分子が飛び立って気体になる数というのは，表面の近くに単位面積および単位時間あたりにどのくらい分子が存在するかに比例し，そこに表面から飛び立つ分子の確率を掛けたものである．数学的に表現すると，

$$N_e = \left(\frac{N}{A}\right)\left(\frac{1}{t_e}\right) e^{-dW/kT} \tag{5.36}$$

となる．液体の場合，分子は密に詰まっているので，単位面積あたりの分子数は単純に A/A_s と書ける．A というのは全面積で，A_s は 1 つの分子に満たされている場所の面積である．単位時間あたりに何個の分子が飛んでいったかを知るために，飛び去るのに要する時間を見積もってみよう．平均速度は v であるということはわかっている．また，分子がごく表面近くのみから飛び出すと仮定する．つまり，分子が進む距離は一層分の距離だけである．もしも一層分の長さが D だとすると，かかる時間は $t_e = D/v$ である．つまり，積

is the area occupied by a molecule. To know how many leave per unit of time, we have to estimate the time it takes to leave. We know the average velocity v, and we suppose that the molecules that leave are the only ones adjacent to the surface, not the ones deep inside the liquid. So the distance they run is only one layer of molecules. If D is the thickness of one molecule layer, the time it takes is $t_e = D/v$. So the product

$$\frac{A}{A_s} \frac{v}{D} \tag{5.37}$$

gives us the number of molecules that leave the area A per unity of time if *all* the molecules were leaving the liquid. This is not true! Only the fraction that has enough energy to win the work function will leave. This fraction is $e^{-dW/kT}$. In conclusion, we fix $A = 1$ (per unity of area) and we have:

$$N_e = \left(\frac{1}{A_s}\right)\left(\frac{v}{D}\right) e^{-dW/kT} \tag{5.38}$$

Now we notice that in situation of equilibrium, the number of molecule that condense N_c (eq. 5.35) and the number that evaporate (eq. 5.38) must be the same. So we have:

$$nv_{gas} = \left(\frac{1}{A_s}\right)\left(\frac{v_{liq}}{D}\right) e^{-dW/kT} \tag{5.39}$$

Now, let's consider finally two things. One is easy: the area A_s of the molecule, multiplied by its thickness D is of course the volume occupied by a molecule $V_a = DA_s$. The other is less intuitive: the velocity v_{gas} of the gas molecules, and the velocity v_{liq} of the liquid *on average* are equal! Why? Because we are in thermal equilibrium. We know that $\frac{1}{2}mv^2 = \frac{1}{2}kT$, so if this was not true, we would have two different average energies and temperatures in the gas and the liquid, and this is not possible. So putting $v_{gas} = v_{liq} = v$ and $V_a = DA_s$ in the equation, we obtain again equation 5.33!

$$nV_a = e^{-dW/kT} \tag{5.40}$$

5.5 蒸発

$$\frac{A}{A_s}\frac{v}{D} \tag{5.37}$$

は面積 A を単位時間あたりに去る分子数を示す．ただしこれはすべての分子が液体から去っていくとした場合の仮定である．この仮定は正しくない．実際にはこの何分の一かが，脱出の仕事関数よりも大きなエネルギーをもって飛んでいく．その割合は $e^{-dW/kT}$ である．まとめると，$A = 1$（単位面積）とすると，

$$N_e = \left(\frac{1}{A_s}\right)\left(\frac{v}{D}\right)e^{-dW/kT} \tag{5.38}$$

とわかる．

ここで平衡状態の条件について気がつくはずである．それは，凝縮する分子の数 N_c(式 5.35) と蒸発する分子の数 (式 5.38) が同じでなければならないということを意味する．すると，

$$nv_{gas} = \left(\frac{1}{A_s}\right)\left(\frac{v_{liq}}{D}\right)e^{-dW/kT} \tag{5.39}$$

となる．

さて，最後に 2 つのことについて考えよう．1 つは簡単なこと．分子の面積 A_s に厚さ D を掛けた量は 1 つの分子によって満たされている体積 $V_a = DA_s$ である．もう 1 つの問題は，直感的にはわかりにくい問題である．それは，気体分子の速度 v_{gas} と，液体の分子の速度 v_{liq} は平均では同じだということである．なぜだろうか？ それは系が熱的に平衡状態にあると考えているからである．私たちは，$\frac{1}{2}mv^2 = \frac{1}{2}kT$ ということを知っている．つまり，これが成り立たない場合には，同じ温度で異なる平均速度の分子が存在することになり，それはありえない．つまり，$v_{gas} = v_{liq} = v$ と $V_a = DA_s$ をこの式に代入すると，私たちは再び式 (5.33) と同様に

$$nV_a = e^{-dW/kT} \tag{5.40}$$

を得るのである．

これは素晴らしい結果である．なぜなら，この結果は，私たちのモデルが活用できるということを示しているからである．式 (5.40) を得るために，は

This is a fantastic results! Why? Because it tell us that our approximate models work. To obtain equation (5.40) in the first case we started from a physical model based only on the ideal gas energies. The second time we started from a molecular mechanical model, using equilibrium and the Boltzmann law concept. We arrive to the same result. This is the power of modelling with statistical mechanics.

Example 5.3

Suppose we have a liquid enclosed in a box in equilibrium with its vapour at room temperature. This system is made of spherical molecules with radius $R = 1$ nm, T is 300 K and molecules are bound together by a potential with work function ϕ of 1 eV. What is the molecular density of the gas state?

We apply equation 5.40:

$$n = \frac{1}{V_a} e^{-dW/kT} \tag{5.41}$$

considering that $dW = 1 \text{ eV} = 1.6 \times 10^{-19}$ J, $k = 1.38 \times 10^{-23}$ J/K, and $T = 300$ K, we have:

$$n = \frac{3}{4\pi R^3} e^{(-1.6 \times 10^{-19}/(1.38 \times 10^{-23} \times 300))}$$

$$= \frac{3}{4\pi \times (10^{-9})^3} e^{(-1.6 \times 10^{-19}/4.14 \times 10^{-21})}$$

$$= 2.39 \times 10^{26} \times 1.64 \times 10^{-17}$$

$$= 3.9 \times 10^9 \; \frac{1}{\text{m}^3} \tag{5.42}$$

this means that, in condition of equilibrium at room temperature, the vapour is made of about one billion molecules per unit of volume against a liquid that is about 6×10^{16} times more dense. What happens if I increase the temperature to 400 K?

じめ，理想気体のエネルギーに基づいた物理モデルからスタートした．2回目は，私たちは平衡とボルツマン則に基づいた分子力学モデルからスタートした．そして，同じ結果にたどり着いた．これは，統計力学のモデル化の力である．

> **例題 5.3**
>
> 箱に閉じ込められた液体について考えてみよう．この液体は室温で蒸気と平衡状態を保っている．この系は，半径 $R = 1\,\mathrm{nm}$ の球状の分子が仕事関数 $\phi = 1\,\mathrm{eV}$ でのポテンシャルにより束縛されている状態にある．気相の分子密度はどのようになっているだろうか．
>
> 式 (5.40) を用いると，
>
> $$n = \frac{1}{V_a} e^{-dW/kT} \tag{5.41}$$
>
> となる．$dW = 1\,\mathrm{eV} = 1.6 \times 10^{-19}\,\mathrm{J}$, $k = 1.38 \times 10^{-23}\,\mathrm{J/K}$, $T = 300\,\mathrm{K}$ を用いると
>
> $$\begin{aligned} n &= \frac{3}{4\pi R^3} e^{(-1.6 \times 10^{-19}/(1.38 \times 10^{-23} \times 300))} \\ &= \frac{3}{4\pi \times (10^{-9})^3} e^{(-1.6 \times 10^{-19}/4.14 \times 10^{-21})} \\ &= 2.39 \times 10^{26} \times 1.64 \times 10^{-17} \\ &= 3.9 \times 10^9 \, \frac{1}{\mathrm{m}^3} \end{aligned} \tag{5.42}$$
>
> となる．この結果は，室温で平衡状態にあるとき，蒸気は単位体積あたり10億個の分子であり，液体は 6×10^{16} 倍の密度になっている．温度が400 K にあがると，どうなるかやってみよう．

Chapter 6 Applications

6.1 Diffusion process

Lets consider the particles of a gas. They move in all direction, and their energy is $\frac{1}{2}kT$ as we know very well. Of course the particles will collide with each other often. How often?

Well, this depends on many things. Of course the particles density is high, we expect to have more collisions. Each particle move straight, undisturbed for a while, then it collides elastically with another particle. The word *elastic* means that the collision preserve the kinetic energy. In other words, the kinetic energy $\frac{1}{2}mv^2$ of the particle before and after the collision remains the same. Let's suppose to have a very big box, with few gas particles in it. Let's suppose that, in this box, each particle has on average one collision every one minute. It's just a supposition. So we can call $\tau = 60$ s the *average time* between collisions of any single gas particle.

Of course, the number of collisions that this particle is experiencing, is depending on time, and *on average* will be equal to

$$N(t) = \frac{t}{\tau} \tag{6.1}$$

This equation is very easy to understand: after a time $t = \tau$, *on average*, the particle will have experienced one collision. If $t = 2\tau$, two collisions and so on.

So let's ask ourselves: what is the probability of having a collision? If we know that every $\tau = 60$ s we have a collision, after only one seconds we have $1/60th$ probability to have one, and after 30 seconds, we will have $1/2 = 50\%$ to have one. This means that dt/τ is the probability to have a collision for one particle.

第6章　応用編：熱力学的な概念を使って

6.1　拡散のプロセス

さて，気体の粒子について考えてみよう．分子はすべての方向に移動し，そのエネルギーはよくご存知のように $\frac{1}{2}kT$ である．もちろん，分子は時々衝突する．どのくらいの頻度で衝突するだろうか．

その頻度は様々な要素に依存する．また，当然ながら密度が高くなると衝突が多くなることが予測できる．それぞれの分子は絶えずまっすぐに動き，ほかの分子と弾性衝突する．この弾性という言葉は，分子が運動エネルギーを失わないということを意味している．他の言葉で言うと，衝突前後での運動エネルギー $\frac{1}{2}mv^2$ は同じだということである．では次に，ごく大きな箱を考えてみよう．この箱には，ごく僅かな気体分子しか入っていない．その箱の中の気体分子は，1分毎に1回衝突するとする．これはただの仮定である．つまり，私たちは平均の衝突時間を $\tau = 60\,\mathrm{s}$ だと仮定しているわけである．τ は粒子が経験する衝突の回数で，時間にもよるが，平均では

$$N(t) = \frac{t}{\tau} \tag{6.1}$$

となる．この式はとても簡単に理解することができる．時間 $t = \tau$ 経過した後，平均すると，分子は1回の衝突を経験している，ということである．もしも時間 $t = 2\tau$ の場合には，衝突経験回数は2回である．

では衝突の確率について考えてみよう．もしも $\tau = 60\,\mathrm{s}$ ごとに衝突することを知っていたとすると，1秒後に1回衝突している確率は $1/60$ になり，30秒後には $1/2 = 50\%$ となることがわかる．これが意味することは，dt/τ というのは1つの分子が衝突する確率というのを示しているということである．

さて，ここで，集団としては何が起こるかを考えてみよう．すべての分子の衝突の確率を考えるとどうなるだろうか．これを理解するために，まずは $N^*(t)$ をまだ衝突したことのない気体分子の数だとしよう．すると，

Now let's think about what's happening collectively. What is the probability that *some* of *all* the particles collide? To understand this, let's define $N^*(t)$ as the number of gas molecules that **did not** collide yet. Then we can write

$$N^*(t+dt) = N^*(t) - N^*(t)\frac{dt}{\tau} \tag{6.2}$$

What is this equation? This equation is very simple. It just means that the particles that do not have hit anything at $t + dt$, is given by the initial particles at t, $N^*(t)$, minus the total collisions occurred in the time dt. The total number of hits during this small interval dt is of course the current number of molecules -again $N^*(t)$- multiplied the probability dt/τ, so we obtain the above eq. (6.2). If we manipulate eq. (6.2) we easily obtain

$$\frac{N^*(t+dt) - N^*(t)}{dt} = -\frac{N^*(t)}{\tau} \tag{6.3}$$

if $dt \to 0$ this becomes

$$\frac{dN^*(t)}{dt} = -\frac{N^*(t)}{\tau}$$
$$\frac{dN^*(t)}{N^*(t)} = -\frac{dt}{\tau} \tag{6.4}$$

and if we integrate we finally have this equation:

$$N^*(t) = N^*(0)e^{-t/\tau} \tag{6.5}$$

This means that the number of particles N^* that do **not** experience any collision, diminish in time with the equation (6.5) above. We can also say that *the probability of* **no** *collision* is $e^{-t/\tau}$ and that the probability of collision is then

$$P(t) \propto 1 - e^{-t/\tau} \tag{6.6}$$

This is the *collective* probability of collision, the probability that *some* of *all* the particles collide. If the time t tends to infinite, this probability of course

6.1 拡散のプロセス

$$N^*(t+dt) = N^*(t) - N^*(t)\frac{dt}{\tau} \tag{6.2}$$

と書くことができる．この式はどんな式だろうか．とても単純で，$t+dt$ において衝突していない粒子の数を意味しており，t での初期の分子の数 $N^*(t)$ から時間 dt の間に起こった全衝突数を引いて得られるものである．dt という短い時間の間に起こる全衝突数は，分子の数（$N^*(t)$）に dt/τ を掛けることで得ることができる．

式 (6.2) を使うと，簡単に

$$\frac{N^*(t+dt) - N^*(t)}{dt} = -\frac{N^*(t)}{\tau} \tag{6.3}$$

を導くことができる．もし $dt \to 0$ ならば

$$\frac{dN^*(t)}{dt} = -\frac{N^*(t)}{\tau}$$
$$\frac{dN^*(t)}{N^*(t)} = -\frac{dt}{\tau} \tag{6.4}$$

となり，積分すると

$$N^*(t) = N^*(0)e^{-t/\tau} \tag{6.5}$$

となる．この式はこれまでに衝突を経験していない粒子の数 N^* が時間とともに減少していることを意味する．または，衝突しない確率が $e^{-t/\tau}$ であり，衝突する確率が

$$P(t) \propto 1 - e^{-t/\tau} \tag{6.6}$$

であるとも言える．これは，**集合としての衝突の確率**であり，衝突した分子**全粒子に対しての衝突する確率**を示している．もし時間 t が無限大になれば，当然この確率は 1 になる．私たちは，平均衝突時間 τ を，**緩和時間**と呼ぶ．なぜなら，これは，平均においてどれだけの時間分子が他の分子に衝突しないで進むことができるかを示しているものだからである．

今，私たちは気体分子がある一定の速度で動いているということを知っている．すべての分子は，それぞれ独自の速さで動いている．そして，それぞれの

tends to one. We can call the average time of collision τ *relaxation time* because it somehow represents -on average- how much time the particle can run without any collision.

Now, we know that our gas particles move with a certain speed. Every particle has its own speed, and at each collision the speed changes. However, we know that -on average- the speed is given by $\frac{1}{2}mv^2 = \frac{1}{2}kT$! If this is true, we can define another important parameter

$$\lambda = \tau v \qquad (6.7)$$

this λ is called usually *mean free path*. It is the average length that the particle can run free. This parameter is as important as τ and has almost the same meaning. The mean distance the gas particle can go -on average- before it experiences a collision.

Now let's try to relate some gas parameters that we know. Let's ask ourselves, what is the probability of collision after the particle has moved a distance dx? Similarly to what we already did when we were considering the *relaxation time*, we know the answer; this probability (for only *one particle*) is:

$$\frac{dx}{\lambda} \qquad (6.8)$$

Now, lets consider it from another point of view. If we have a box of volume dV, with particle density n, what is the probability of collision in a small volume of area A and length dx? Well, as we know the number of particles in this section is $nAdx$, (see figure 6.1).

If every single particle has an average area σ, then the total area occupied by the particle is $\sigma nAdx$. If we divide this area with the total area available (which of course is A), the we obtain the probability of collision within this section of length dx: $\sigma n dx$. But this value must be equal to eq. (6.8), so

衝突において，その速度を変化させる．しかし，私たちは $\frac{1}{2}mv^2 = \frac{1}{2}kT$ によって平均の速度を知ることができる．もしもこれが本当なら，私たちはもう1つ重要なパラメータを定義できる．

$$\lambda = \tau v \tag{6.7}$$

この λ は，**平均自由行程**と呼ぶ．これは，気体分子がほかの分子と衝突しないで動くことができる平均の距離を示している．このパラメータは τ と同じ様に重要で，同じような意味合いを持っている．この平均自由行程分だけ，分子は衝突せずに自由に進むことができるということである．

では，いくつかの気体分子に関するパラメータについて，それぞれの関係を明らかにしよう．分子が dx だけ動いた後に衝突する確率はどうなっているだろうか．これまで緩和時間を導くためにやってきたのと同様のことを考えればいい．私たちはすでに答えを知っているのである．この確率は（**1つの粒子に**とってみると），

$$\frac{dx}{\lambda} \tag{6.8}$$

となる．

ここで違う観点から考えてみよう．もしも体積 dV の箱に気体分子密度が n だけ詰められていたとすると，面積 A で長さ dx の小さな体積内で衝突する確率はどうなっているのだろう？　この領域につまっている分子の総数は，$nAdx$ である．（図 6.1 参照）．

もし，1つの分子の衝突断面積が σ だとすると，すべての分子によって占められる総面積は $\sigma nAdx$ となる．分子の占める総面積 $\sigma nAdx$ をこの領域の断面積 A で割ると，結果は距離 dx の間に衝突する単位面積あたりの確率 $\sigma n dx$ となる．この値は式 (6.8) と同じでなければならないので，

$$\frac{dx}{\lambda} = \sigma n dx \tag{6.9}$$

となる．式を整理すると，

$$n\sigma\lambda = 1 \tag{6.10}$$

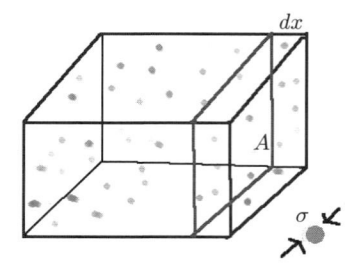

Fig. 6.1 A scheme for the modelling of the collision probability. The *cross section* is represented by the average area of a single gas molecule.

finally we have:

$$\frac{dx}{\lambda} = \sigma n dx \tag{6.9}$$

simplifying this yields this very important relation:

$$n\sigma\lambda = 1 \tag{6.10}$$

this equation is very important because it relates the *particle density n*, their *cross section* σ and their *mean free path* λ in a gas system each other with a very simple and compact formula.

Example 6.1

Suppose you have a system of particles of radius 1 nm that move in space with an average collision distance of 1 μm (1 thousand times the radius). What is the particle density n? With eq. (6.10) we can immediately estimate:

$$n = \frac{1}{\sigma\lambda}$$

if the radius is 1 nm, a good estimation of the cross section for a spherical particle is $\sigma = 4\pi R^2 = 4\pi(10^{-18})$ m^2. The *mean free path* in this case is $\lambda = 10^{-6}$ m, so finally we have:

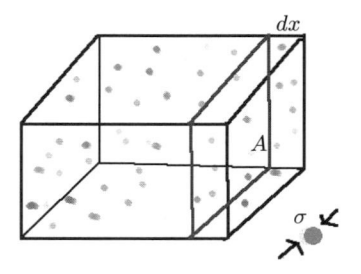

図 6.1 衝突確率を考えるためのモデルの模式図. 断面積は 1 つの分子がもつ平均面積によって表される.

を得ることができる. この式は非常に大切である. なぜなら, これは, 分子密度 n と, その断面積 σ, そして平均自由行程 λ が非常にコンパクトにその関係を示しているものだからである.

> **例題 6.1**
>
> 半径 1 nm の粒子が平均自由行程 1 μm（半径の 1000 倍）で存在している系の密度 n について考えよう. 粒子の密度は式 (6.10) より, すぐに見積もることができる.
>
> $$n = \frac{1}{\sigma \lambda}$$
>
> 半径は 1 nm なので, 球状の粒子の衝突断面積は $\sigma = 4\pi R^2 = 4\pi (10^{-18})$ m^2 である. 平均自由行程は $\lambda = 10^{-6}$ m なので, 最終的に密度を
>
> $$n = \frac{1}{4\pi \times 10^{-18} \times 10^{-6}} \approx 10^{23} \left[\frac{1}{\text{m}^3}\right]$$

$$n = \frac{1}{4\pi 10^{-18} \times 10^{-6}} \approx 10^{23} \left[\frac{1}{m^3}\right]$$

6.2 The drift velocity

Now suppose we have some particles in the middle of other particles of another kind. They can be gas particles in the middle of other heavier molecules. Or they can be electrons in a metal, between atomic ions. Suppose these particles are subject to a force, let's call it F. Let's suppose that this force will push the particles in one specific direction, but they will not influence the other kind of particles. As we know, for a small interval of time, the particles will accelerate freely under the force F. Then there will be a collision, maybe with the other kind of particles. Now, during the free time they are solely subject to the force F, they will accelerate. How much? Of course the acceleration will be

$$a_p = \frac{F}{m} \tag{6.11}$$

where m is the mass of the particle.

After how much time will there be a collision? We do not know exactly. However, we already defined the *average* time between collisions. We called this time *relaxation time* and used the symbol τ. If we suppose that after every collisions our speed is reset and we have to start again, then we can calculate how much is the *average* speed that the particles move. Of course it is the average acceleration the particles have, multiplied the average time they move freely in space. It is an *average, collective* speed, and usually is called *drift* speed:

$$v_{drift} = a_p \tau = \frac{F\tau}{m} \tag{6.12}$$

In general the velocity of drift is expressed under the following form:

と見積もれる.

6.2 ドリフト速度

さて，ここで，いくつかの粒子が違う種類の他の粒子と混ざっている場合について考えよう．この粒子というのは，重さが違う数種類の分子が混在していたり，あるいは金属の原子イオンの中にある電子だと考えることもできる．そのような粒子に働く力 F を考えてみよう．この力というのは，ある一方向に粒子を動かすものだと考える．ただし，他の粒子には影響がないとする．ご存知のように，短い時間の間では粒子は力 F のもとで自由に加速する．しばらくすると衝突が起こるが，その衝突はほかの種類の粒子と起こるだろう．どのくらいの時間，粒子は F を受けて自由に加速できるだろうか．加速度は

$$a_p = \frac{F}{m} \tag{6.11}$$

となる．m は粒子の質量である．

どれだけ時間がたてば衝突が起こるのか，そのものを知ることはできない．しかし，私たちは衝突と衝突の間の平均時間を定義し，これを，**緩和時間**と呼んで τ という記号を使っている．仮に，衝突ごとに速度がリセットされ，もう一度加速し始めると仮定すると，粒子が動く平均速度を求めることができる．粒子の平均の加速度と自由に空間を動き回れる平均時間を掛け合わせると，粒子の平均の速度が求まる．この粒子の平均速度はドリフト速度と呼ばれ，

$$v_{drift} = a_p \tau = \frac{F\tau}{m} \tag{6.12}$$

となる．

通常，ドリフト速度は次の式から求めることができる．

$$v_{drift} = \mu F \tag{6.13}$$

μ は非常に重要なパラメータで，**移動度**と呼ばれる．移動度は

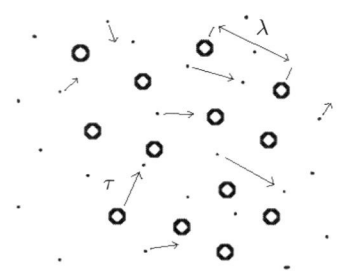

Fig. 6.2 A scheme describing the drift process. Smaller particles in movement under a force F collide against heavier ones supposed fixed. The average relaxation time is τ and the average *mean free time* is λ. If these parameter are true on average, it results equation 6.12.

$$v_{drift} = \mu F \qquad (6.13)$$

The parameter μ is very important, called *mobility* and it is expressed as

$$\mu = \frac{\tau}{m} \qquad (6.14)$$

We have to remember that drift velocity for a system of particles can be defined only between some kind of particles against some other. We cannot define drift if we have only one species of particles in our box (!)

Example 6.2

We open a text book and we find that the mobility of silicon at room temperature is $\mu = 1400 \text{ cm}^2/\text{Vs}$. What does it means and what is the average τ in this case?

First of all let's check the dimensions of this μ relation in MKS units, $1400 \text{ cm}^2/\text{Vs} = 0.14 \text{ m}^2/\text{Vs}$, remembering that Volts in MKS is $\text{m}^2\text{kg}/\text{s}^3\text{A}$, using [ ] brackets to indicate units, we have:

$$\mu = \frac{[m^2(s^3 A)]}{[s(m^2 kg)]} = \left(\frac{[s]}{[kg]}\right)([sA])$$

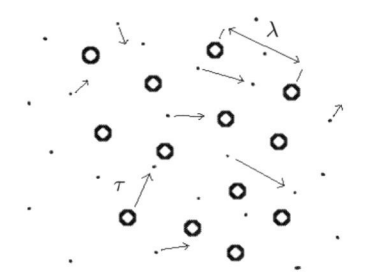

図 6.2 ドリフト過程の模式図. 小さな粒子が力 F で動き, 重たい粒子と衝突する. 重たい粒子は固定していると考える. 平均の緩和時間は τ, 平均自由行程は λ で表す. もしも平均において, このパラメータが本当ならば, 式 6.12 が導かれる.

$$\mu = \frac{\tau}{m} \tag{6.14}$$

という式で示される. ただしドリフト速度というものが, 異なった粒子の間だけに定義されるということを忘れてはいけない. もしも 1 つの箱に 1 種類の粒子しかなかったら, ドリフト速度を定義することはできない.

例題 6.2

教科書を開くと, 室温でのシリコンの移動度が $\mu = 1400\,\mathrm{cm^2/Vs}$ と載っていた. この意味と平均の τ を考えよう.

まず, この μ が MKS 単位系だとどうなるか確認してみよう. 1400 $\mathrm{cm^2/Vs} = 0.14\,\mathrm{m^2/Vs}$ となる. V は MKS 単位系では $\mathrm{m^2 kg/s^3 A}$ となる. [ ] を使って単位を示してみると,

$$\mu = \frac{[\mathrm{m^2(s^3 A)}]}{[\mathrm{s(m^2 kg)}]} = \left(\frac{[\mathrm{s}]}{[\mathrm{kg}]}\right)([\mathrm{sA}])$$

となる. μ は式 (6.14) によって表される物理量なので, $\mu = [\mathrm{s}]/[\mathrm{kg}]$ とな

We immediately notice that our physical definition of μ is instead given by eq. (6.14), $\mu = [\text{s}]/[\text{kg}]$. We have this difference in definition because in electronics, people define the drift velocity v_d as $v_d = \mu E$ not $v_{drift} = \mu F$ as we do in physics. But the two definition just coincide if we say that we have a unit of charge, in fact $F = Eq$ where q is the *test charge* in the electric field. All as we know from high school! Since the charge is expressed in MKS as Ampere multiplied by seconds ([As]) the two definitions coincide.

Coming back to our problem, what is the average relaxation time? If the force act on a single electron charge, then:

$$\begin{aligned} \tau &= \mu m_e \\ &= \left(0.14 \frac{[\text{s}^2\text{A}]}{[\text{kg}]}\right)(9 \times 10^{-31}[\text{kg}]) \approx 1.3 \times 10^{-31}[\text{s}]([\text{As}]) \\ &= 1.3 \times 10^{-31}[\text{s}][\text{C}] \end{aligned} \qquad (6.15)$$

this value is intended for unit of charge, so the [C] goes away and our τ is expressed in seconds.

6.3 Electric resistance

Let's now apply what we know to a conduction problem. Let's suppose we have a mixture of gas between two electrodes, like in figure 6.3. The drifting particles are moving under a force. The other particles can be neutral atoms that do not move and oppose the drifting process. The two electrodes are under a potential V. As we know from physics this will correspond to a electric field of value $E = V/d$, where d is the separation of the two electrodes. What is the force acting on these ions? If the charge of the ion is q, from basic physics we know it is $F = qE$. So using eq (6.13) we have:

る．この違いというのは，電子工学ではドリフト速度 v_d を物理学の世界で使う $v_{drift} = \mu F$ ではなく $v_d = \mu E$ と定義していることによる．定義は異なるが，電荷が 1 のときに，q が電場中の試験電荷だとすると，$F = Eq$ となる．これは高校のときに勉強した通りである．電荷は MKS 単位系では時間とアンペア数を掛けた As で示されるので，2 つの定義が一致するのである．

話題を元に戻そう．平均緩和時間はどうなるだろうか．仮に 1 つの電荷に力が働くとすると，

$$\begin{aligned}
\tau &= \mu m_e \\
&= \left(0.14 \frac{[\text{s}^2\text{A}]}{[\text{kg}]}\right)(9 \times 10^{-31}[\text{kg}]) \approx 1.3 \times 10^{-31}[\text{s}]([\text{As}]) \\
&= 1.3 \times 10^{-31}[\text{s}][\text{C}]
\end{aligned} \tag{6.15}$$

となる．この値は電荷の単位になっている．そのため [C] はなくなり，τ は秒で表される．

6.3 電気抵抗

さて，続いては，伝導の問題に取り組もう．図 6.3 のように，2 つの電極の間に，混じり合った気体分子があると想像しよう．ドリフトする粒子には，ある力が動いている．もう一方の粒子は中性の原子で，動かずにドリフト過程を妨害する．2 つの電極の間には V というポテンシャルがあり，電場 $E = V/d$ を生じる．d は，2 つの電極間の距離である．イオンに働く力は何だろうか．もしもイオンの電荷が q ならば，物理の基礎的な式から，力は $F = qE$ であるとわかる．つまり，式 (6.13) から，

$$v_{drift} = \mu F = \mu q E = \mu q \frac{V}{d} \tag{6.16}$$

という式を導くことができる．

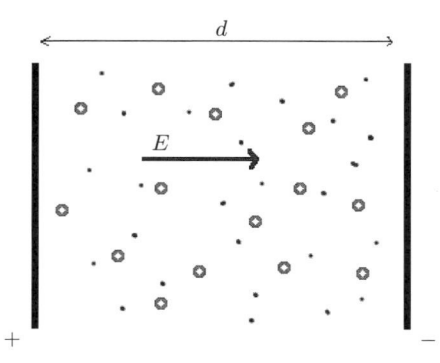

Fig. 6.3 A scheme describing the conduction process. The bigger particles are those which are fixed and oppose the drift process, neutral atoms for example. The smaller particles represent the drifting ions.

$$v_{drift} = \mu F = \mu q E = \mu q \frac{V}{d} \qquad (6.16)$$

Now let's make another physical model of the charge movement in the gas. We suppose -as we did many times- that the density of drifting particles is known and its value is n_i, we give the index i because in this case the moving particles are ions. What is the electric current at the electrodes? Current is defined in physics as the number of charges arriving at the electrode in a time t per unit of area A. We ask ourselves: how many ions per unit of time are arriving at the electrode, if the velocity is v_{drift}? In time t the particles move on average for a distance $v_{drift} t$. If we multiply this small distance for the area of the electrodes A_e we have the volume containing all the ions that will reach the electrode within the time t. This volume multiplied by the density n_i will result in the total number N_e of ions reaching the electrode in a time t.

$$N_e = n_i A_e v_{drift} t \qquad (6.17)$$

Now let's come back to our question: what is the electric current at the electrodes? Well, the current is the number of charges per second per unit

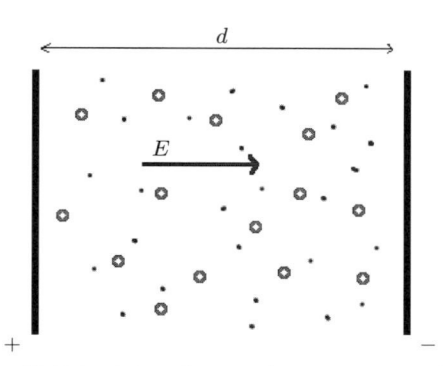

図 6.3 伝導プロセスの模式図. 大きな分子は固定されていて, 例えば中性原子などを想像するといい. 小さな粒子はドリフトイオンである.

さて, 電荷の動きに関する物理モデルについて考えてみよう. 私たちはまずこれまで何度も行ったのと同様にドリフト粒子の密度が n_i であると仮定する. i は動いている粒子がイオンだということを示している. 電極での電流はどうなるだろうか. 物理的には電流は単位時間 t に単位面積 A にやってくる電荷の数として定義される. それをふまえて自問してみよう. ドリフト速度が v_{drift} のとき, 単位時間にどのくらいのイオンが電極に到達するだろうか. 時間 t の間に粒子が平均的に動く距離は $v_{drift}t$ である. 電極の面積 A_e にこの短い距離をかけると, 時間 t の間に電極にあたるすべてのイオンを含む体積を計算することができる. この体積にイオン密度 n_i を掛け合わせることによって, 時間 t の間に電極に到達するイオンのすべての数 N_e がわかる. つまり,

$$N_e = n_i A_e v_{drift} t \tag{6.17}$$

となる.

ここで, 電極に流れる電流はどのくらいだろうという最初の疑問に戻ろう. 電流というのは, 単位時間あたり, 単位面積あたりの電荷の数である. つま

of area. So if multiply by the charge of every single ion q and divide by the unit of area and by the time t, we have what we want:

$$I = qn_i A_e v_{drift} \tag{6.18}$$

Now if we substitute the value of v_{drift} we have:

$$I = \frac{\mu q^2 n_i A_e}{d} V \tag{6.19}$$

(Please remember that to divide by the unit of area is like to divide by one, as long as the other area A_e, the area of the electrodes, is expressed in that unit). Now you notice that what we just found is nothing else than the Ohm law $V = IR$ (!). We immediately find the expression of Resistance R for an ionized gas under an electric field

$$R = \frac{d}{\mu q^2 n_i A_e} \tag{6.20}$$

Also, if we remember from any textbook the general expression of the resistance as $R = \rho l/S$ where l is the distance of the electrodes, S their area and ρ is the so call *resistivity*, then we can also find an expression of the resistivity in terms of molecular parameters:

$$\rho = \frac{1}{\mu q^2 n_i} \tag{6.21}$$

Example 6.3

Suppose that you have two electrodes of 10 squared centimeters placed 1 centimeter apart, these are immersed in a gas with mobility $\mu = 10^5$ s/kg and density $n_i = 10^{30}$ particles per unit of volume. If the ion has one unit charge only, what will be the resistance R of the gas?

To do this calculation lets first verify that the unities are OK:

$$R = \frac{d}{\mu q^2 n_i A_e} = \frac{[\text{m}]}{[\frac{\text{s}}{\text{kg}}][\text{A}^2\text{s}^2][\frac{1}{\text{m}^3}][\text{m}^2]}$$

6.3 電気抵抗

り，個々のイオンの電荷 q を掛け，単位面積と時間 t で割ると，

$$I = qn_i A_e v_{drift} \tag{6.18}$$

となる．ドリフト速度の値を v_{drift} を代入すると，

$$I = \frac{\mu q^2 n_i A_e}{d} V \tag{6.19}$$

となる（注：単位面積で割るということは，1で割ることと同じである．ただし，そのときには単位系を揃えないといけない）．ここでオームの法則 $V = RI$ を使うと，直ちに電場におけるイオン気体の抵抗値 R を求めることができ，下のようになる．

$$R = \frac{d}{\mu q^2 n_i A_e} \tag{6.20}$$

また，もしも $R = \rho l / S$ という一般的な教科書にかいている式を覚えているとすると，（l は電極間距離，S はその面積，ρ は抵抗率）分子のパラメータで抵抗率を知ることができるのである．

$$\rho = \frac{1}{\mu q^2 n_i} \tag{6.21}$$

例題 6.3

$10\,\mathrm{cm}^2$ の2つの電極が $1\,\mathrm{cm}$ 離れてあるという状況について考えてみよう．これらは移動度 $\mu = 10^5\,\mathrm{s/kg}$，単位体積あたりの密度 $n_i = 10^{30}$ 個の気体中にある．もしもイオンが1つの電荷だけもっているとすると，気体の抵抗値 R はどうなるだろうか．

この計算をするにあたり，まずはその単位について確認してみよう．

$$R = \frac{d}{\mu q^2 n_i A_e} = \frac{[\mathrm{m}]}{[\frac{\mathrm{s}}{\mathrm{kg}}][\mathrm{A}^2\mathrm{s}^2][\frac{1}{\mathrm{m}^3}][\mathrm{m}^2]}$$

that is:

$$R = \frac{[m]^2[kg]}{[s]^3[A]^2}$$

and this is exactly the definition of Ohm, so we are indeed OK with units! Now, let's substitute our values and find the resistance in this example:

$$R = \frac{d}{\mu q^2 n_i A_e} = \frac{10^{-2}}{(10^5)(1.6 \times 10^{-19})^2(10^{30})(10^{-3})} = 3900 \, \Omega$$

and this is the resistance we were looking for.

6.4 Diffusion

What is the difference between *drift* and *diffusion*? *Diffusion* is a natural movement of particles that do not require an external force. Let's firstly define diffusion. You remember the definition of current? We have an electric current if a certain amount of electric charges passes a unit of area in a certain time. Now, we can define another current, we call it *molecular current*, it is defined by the number of molecules that flow in a unit of area per unit of time, very similar to the definition of electric current. If we know the density of molecules n_a, we can easily calculate the total number of molecules in the time t, as we did usually:

$$N_{tot} = n_a dV = n_a A v_x t \tag{6.22}$$

with dV representing the small volume of particles moving in the time t, A the section area we are considering and v_x the average velocity of the molecules in the direction x. So we can write that the molecular current is

$$J_x = n_a v_x \tag{6.23}$$

This molecular current exists because of a force pushing the molecules around. But we do not have this force in this case, otherwise we would

すなわち，

$$R = \frac{[\text{m}]^2[\text{kg}]}{[\text{s}]^3[\text{A}]^2}$$

これは Ω の定義と同じである．次に，最初に考えた設定の数字を入れて，計算してみよう．すると求めている結果が下記のように得られる．

$$R = \frac{d}{\mu q^2 n_i A_e} = \frac{10^{-2}}{(10^5)(1.6 \times 10^{-19})^2(10^{30})(10^{-3})} = 3900\,\Omega$$

6.4 拡散

拡散とドリフトの違いというのは何だろうか．拡散というのは，外力を必要としない粒子の自然な動きによるものである．まず，拡散について定義しよう．電流の定義を覚えているだろうか．電流は単位面積あたり単位時間に通過する電荷の総量である．では，もう 1 つの電流について考えよう．それを分子電流と呼ぼう．これは，単位面積あたり，単位時間あたりに流れる分子の数と定義する．電流の定義に似ている．もしも分子密度 n_a を知っているとすると，時間 t の間の分子の数は簡単に

$$N_{tot} = n_a dV = n_a A v_x t \tag{6.22}$$

という式を使って計算できる．ここで dV は時間 t の間に分子が動き回る微小な体積であり，A は断面積，v_x は x 方向の分子の平均速度である．つまり，私たちは分子電流をこう書き換えることができる．

$$J = n_a v_x \tag{6.23}$$

この分子電流は，分子を押す力がある場合には，本当に存在するかもしれない．しかし，今考えている場合は力は働いておらず，違う方法でドリフトについて語らなくてはいけない．

まず，どんな力も働いていないのに分子が動き回る理由を考える必要があ

be talking about *drift*.

So we need to have a reason for the molecules to move around, without any force. This reason must be the difference in concentration! So we have as usual to *think small*. Let's imagine that there are two concentrations, n_- and n_+. These are the concentration of the molecules on the left and right of an imaginary x axis **before** the molecules begin to move.

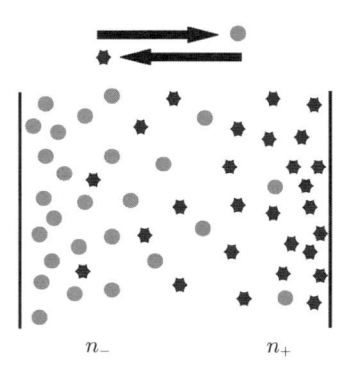

Fig. 6.4 A scheme describing the diffusion process of two different species at different concentration. Note that there is not an external force F in this case(!)

In this situation we have two concentrations, n_+ and n_- representing the concentration of molecules near an imaginary separation axis. Let's consider the difference (the *differential*) between these two concentrations $dn_a = (n_+ - n_-)$. We can define the molecular current $J_x = dn_a v_x$, and for easy mathematical reasons

$$dn_a = \frac{dn_a}{dx}\Delta x = \frac{dn_a}{dx}\lambda \qquad (6.24)$$

where λ is the free mean path. Why we chose λ? Because we are *thinking small* we use the smallest possible movement before collisions stop, the free mean path. So our current is

$$J_x = -\lambda v_x \frac{dn_a}{dx} \qquad (6.25)$$

る．実はこの理由は，濃度の差によるものである．まずは小さい系から考え始めよう．2種類の濃度 n_- と n_+ をもつ系を想像しよう．これらの濃度の状態の分子が，動き始める前には x 軸に沿って左と右にあるとする．

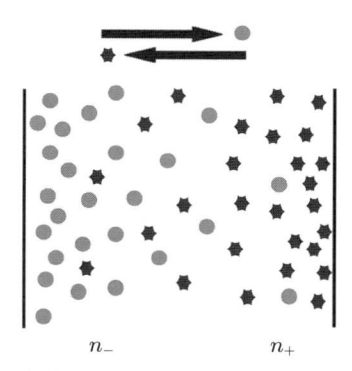

図 6.4 2種類の違う種が違う濃度にあるときの拡散についての概念図．ここで，外部の力 F が存在しないことに注目してほしい．

この状況では，n_+ と n_- は距離 x 離れた2つの濃度を仮定している．それでは，2つの濃度の差 $dn_a = (n_+ - n_-)$ について考えてみよう．私たちは数学的な理由から，$J_x = dn_a v_x$ と定義した．

$$dn_a = \frac{dn_a}{dx} \Delta x = \frac{dn_a}{dx} \lambda \tag{6.24}$$

ここで λ は平均自由行程のことである．どうして λ を Δx として選んだのだろうか．それは，私たちが小さいスケールでの出来事を考えているので，分子の衝突が起こる前に動くことができる距離，すなわち平均自由行程を考えることにしたからである．つまり，今考えている電流は，

$$J_x = -\lambda v_x \frac{dn_a}{dx} \tag{6.25}$$

となり，これはすべての xyz 方向で一般的に成り立つ．マイナスの符号は，

this is valid in general for every other xyz direction. The sign $-$ is because J_x runs opposite to the concentration difference. We remember that $\lambda = v_x \tau$ and $\tau = \mu m$ then:

$$J_x = -v_x^2 \mu m \frac{dn_a}{dx} \tag{6.26}$$

Let's recall the fundamental equation $\frac{1}{2}mv_x^2 = \frac{1}{2}kT$ and we have

$$J_x = -\mu kT \frac{dn_a}{dx} \tag{6.27}$$

In conclusion, if we define the *diffusion coefficient* D, we have

$$J_x = -D \frac{dn_a}{dx} \tag{6.28}$$

where $D = \mu kT$.

Now, what will happen if we have a combination of drift and diffusion? In this case the drift velocity should compensate for the diffusion, so $J_x = -n_a v_{drift}$ (the minus appears because the two velocities are in the opposite direction).

Knowing that $v_{drift} = \mu F$, we substitute on the above eq. (6.28) and we have

$$D \frac{dn_a}{dx} = n_a \mu F \tag{6.29}$$

and because $D = \mu kT$ we have finally

$$\frac{dn_a}{dx} = \frac{n_a F}{kT} \tag{6.30}$$

which is a very important relation that was first found by A. Einstein[33]. This relation shows that when we have an external force F, the combined effect of diffusion and drift provoke a *gradient* of concentration equal to $n_a F/kT$ as in the equation (6.30). This is valid -of course- only at equilibrium. There is something very important related to equation (6.30). There

---

33) A. Einstein: アルバート・アインシュタイン．有名な物理学者の綴りは覚えておくとよい．

6.4 拡散

J_x が濃度勾配の逆向きに流れるからである. $\lambda = v_x \tau$ と $\tau = \mu m$ より,

$$J_x = -v_x^2 \mu m \frac{dn_a}{dx} \tag{6.26}$$

を導くことができる. 基礎的な式, $\frac{1}{2}mv_x^2 = \frac{1}{2}kT$ を思い出すと,

$$J_x = -\mu kT \frac{dn_a}{dx} \tag{6.27}$$

となる.

結論をいうと, もしも私たちが拡散係数 D を

$$J_x = -D \frac{dn_a}{dx} \tag{6.28}$$

として定義すると, $D = \mu kT$ という関係を得ることができる.

今, ドリフトと拡散を併せて考えると, どういうことがわかるだろうか. この場合, ドリフト速度は拡散を保障するべきである. つまり, $J_x = -n_a v_{drift}$ となる (マイナスが出てくるのは, 2つの速度が反対方向なため).

ドリフト速度は $v_{drift} = \mu F$ だとわかっているので, これを式 (6.28) に代入すると,

$$D \frac{dn_a}{dx} = n_a \mu F \tag{6.29}$$

という式を得る. そして, $D = \mu kT$ という関係から, 最後に私たちは

$$\frac{dn_a}{dx} = \frac{n_a F}{kT} \tag{6.30}$$

という関係を得る. この式はアインシュタインによって明らかにされたとても重要な関係である. この関係は, 外部の力 F があるときに, 拡散とドリフトの効果が合わさることで, 式 (6.30) で表される $n_a F/kT$ と一致する濃度の勾配を引き起こすことを示している. これはもちろん, 平衡状態でのみ当てはまる. 式 (6.30) に関係したとても重要なことがある. この式には2つの変数, n_a (粒子の濃度) と, x (空間的な変数) がある. これらの変数を移項すると,

$$\frac{dn_a}{n_a} = \frac{F dx}{kT} \tag{6.31}$$

are two variables, n_a the particles concentration and x the spatial variable. Let's put these variables together:

$$\frac{dn_a}{n_a} = \frac{F dx}{kT} \qquad (6.31)$$

Interestingly, $F dx$ is the *work* done on the molecules. Now we can integrate the two sides:

$$\int \frac{dn_a}{n_a} = \frac{1}{kT} \int F dx \qquad (6.32)$$

We remember from basic physics that $F = -dU/dx$, which means that the potential energy is $U = -\int F dx$, so:

$$\ln \frac{n_a}{n_0} = \frac{1}{-kT} U \qquad (6.33)$$

Let's forget the index a in the gas-concentration parameter n, so for simplicity for now on $n = n_a$, applying the exponential function left and right, we have:

$$\frac{n}{n_0} = e^{-\frac{U}{kT}} \qquad (6.34)$$

we can write finally this surprising equation:

$$n = n_0 e^{-\frac{U}{kT}} \qquad (6.35)$$

What's so exiting now about this equation? It is exiting the fact that we already know it, it is exactly equation (4.19), the equation for *ideal gas* we found many sections above...! Remember that all our modelling where somehow approximate. We used many simplifications and assumption, however using this approach we reached equation (6.30) studying the effect of diffusion and drift on particles and we discovered that this equation contains in itself the ideal gas relations we started from. This means all our reasoning make sense and that there are no contradictions in the treatment.

6.4 拡散

となる．Fdx は分子のした仕事である．両辺を積分すると，

$$\int \frac{dn_a}{n_a} = \frac{1}{kT} \int F dx \tag{6.32}$$

となる．私たちは，物理の基本的な法則，$F = -dU/dx$ を覚えている．U はポテンシャルエネルギーである．これが意味することは，$U = -\int F dx$ なので，

$$\ln \frac{n_a}{n_0} = \frac{1}{-kT} U \tag{6.33}$$

となる．ここで，気体の濃度 n の引数 a を忘れてしまおう．すると，$n = n_a$ と簡単になり，両辺の指数をとると

$$\frac{n}{n_0} = e^{-\frac{U}{kT}} \tag{6.34}$$

となり，最後に

$$n = n_0 e^{-\frac{U}{kT}} \tag{6.35}$$

という式にたどり着くのである．

　この式に到達するとはとても興奮すべきことである．なぜならこの事実を，私たちはもう知っているからである．この式こそ式 (4.19) で理想気体の式として，すでに前に何度も勉強してきたものである．これはとても興奮すべき事柄である．なぜなら，私たちは，近似を用いたモデル化を行ってきて，様々な場面で簡易化を行い，仮定を使ってきた．しかし，それでも粒子の拡散効果やドリフトについて，理想気体の関係を含んでいる式 (6.30) に到達したのである．つまり，これが意味することとは，すべての取り扱いに，矛盾がなかったということである．

Example 6.4

Let's see in real life what are the numbers involved in equation (6.30). Suppose we have again a ionized gas in an electrode of concentration 10^{23} molecules per cubic meters, but this time the gas molecules *drift* in another gas that is *diffusing* on the opposite direction. If the electrodes are 10 square centimetres, separated by one centimetres with a potential of 1 Volt, what is the gradient concentration at room temperature $T=300$ K? We suppose again that the ionization is of valence one, so $q = 1.6 \times 10^{-19}$, we have:

$$\frac{dn_a}{dx} = \frac{n_a F}{kT} = \frac{n_a(qE)}{kT}$$

we know that with the electrodes separated by 1cm, the electric field E must be $E = 100$ V/m. Then

$$\frac{dn_a}{dx} = \frac{n_a(qE)}{kT} = \frac{10^{23} \times (1.6 \times 10^{-19} \times 10^2)}{(1.38 \times 10^{-23})(300)} \approx \frac{1.6 \times 10^6}{4 \times 10^{-21}}$$
$$= 4 \times 10^{26} \frac{[\text{molecues}]}{[\text{m}^4]}$$

We indicated again with square brackets the units. Considering that the electric field E is [mkg s^{-3}A^{-1}] and that the charge q is [As], you can verify yourself that the units are OK.

6.5 Black body radiation

Suppose we want to study the radiation emission of a hot body. In a hot body, all the molecules vibrates. We can approximate these vibrations as elastic oscillations of a mass particle. If there is radiation emission, it means this particle is charged. It is known to physicists that a charged particle oscillating radiate away an amount of energy W proportional to the energy of the particle, accordingly to this formula:

例題 6.4

式 (6.30) が意味する数がどのようなものであるか，現実世界で考えてみよう．濃度 10^{23} /m^3 のイオン化した気体が電極間にあるとしよう．気体分子は反対方向へと拡散している他の気体中をドリフトしている．電極が $10\,\mathrm{cm}^2$，電極間距離が $1\,\mathrm{cm}$，ポテンシャルが $1\,\mathrm{V}$ だとすると，室温 ($300\,\mathrm{K}$) での濃度の勾配はどうなっているか考えよう．ただしイオンはすべて一価だとする．電荷は $q = 1.6 \times 10^{-19}$ [C] なので，

$$\frac{dn_a}{dx} = \frac{n_a F}{kT} = \frac{n_a(qE)}{kT}$$

となる．電極間距離は $1\,\mathrm{cm}$ なので，電場 E は $100\,\mathrm{V/m}$ となる．つまり，

$$\frac{dn_a}{dx} = \frac{n_a(qE)}{kT} = \frac{10^{23} \times (1.6 \times 10^{-19} \times 10^2)}{(1.38 \times 10^{-23})(300)} \approx \frac{1.6 \times 10^6}{4 \times 10^{-21}}$$
$$= 4 \times 10^{26} \frac{[\text{molecues}]}{[\text{m}^4]}$$

となる．[ ] 内は単位を示している．電場 E は $[\mathrm{mkg\ s^{-3}A^{-1}}]$，電荷 q は $[\mathrm{C}]=[\mathrm{As}]$ なので，単位で計算が正しいかどうか，確認できる．

6.5 黒体放射

熱い物体からの電磁波の放射について考えてみよう．熱い物体では，すべての分子は激しく振動している．この振動について，質量をもった粒子が弾性振動しているとみなして取り扱ってみよう．もしも電磁波の放射があるとすると，それは粒子が帯電しているということである．物理学者には知られていることだが，帯電した粒子は振動すると電磁波を放射し，その強さは粒子のエネルギー W に比例して，以下の式で表される．

6.5 Black body radiation

$$I = \frac{dW}{dt} \propto \gamma W \tag{6.36}$$

This means that the energy radiated away per unit of time is proportional to the energy of the charge with a coefficient γ. On average, at a certain temperature T the energy of the particle is known to be kT, so we have:

$$\frac{dW}{dt} \propto \gamma kT \tag{6.37}$$

We do not derive here the detail of the calculations, however γ is ω/Q where Q is the oscillation *quality factor*[34]. This quality factor is proportional to $1/\omega$. If so

$$\gamma \propto \omega^2$$

Finally we find in the relation for the intensity $I(\omega)$ a direct dependence with ω^2 and the molecules kinetic energy: kT,

$$I(\omega) \propto \omega^2 K_B T \tag{6.38}$$

this equation is called the *Rayleigh's law* for the *black body* radiation. Interestingly enough, this equation fits very well the experimental data for low energies, however but it fail completely to predict a diminished emission at higher energies. Accordingly to eq. (6.38), because of the ω^2, we should observe a lot of X-Ray and other unhealthy emission from a hot body, much more UV and X-rays than other radiation! Instead, of course, in the experiments, we never observe emission over UV or X-rays, even if we heat-up a body at very high temperatures ! Even if it looks correct, there is something completely wrong in this equation, what is going on here ?!? Many researchers at that time studied the problem, and nobody was able to find an answer. The problem was called by physicists of the time *the UV catastrophe*.

---

[34] quality factor(Q 値)：共鳴の鋭さを表す量．振動・波動の教科書の強制振動の章を参照のこと．

$$I = \frac{dW}{dt} \propto \gamma W \tag{6.36}$$

これは，単位時間あたりに放出するエネルギーが，帯電した粒子のエネルギーに比例定数 γ で比例するということを示している．平均では，ある温度 T における粒子のエネルギーは kT として知られているので，

$$\frac{dW}{dt} \propto \gamma kT \tag{6.37}$$

と書き直すことができる．

ここでは，詳しい計算の経緯は省く．しかし，γ は振動の Q 値を Q と書くと，ω/Q である．さらに Q 値は $1/\omega$ に比例するので，

$$\gamma \propto \omega^2$$

となる．すると強度 $I(\omega)$ が ω^2 に直接依存するということがわかる．分子の運動エネルギーは kT なので，

$$I(\omega) \propto \omega^2 kT \tag{6.38}$$

となる．この式は黒体放射を示すレイリーの法則あるいはレイリー・ジーンズの放射法則と呼ばれるものである．面白いことに，この式は低いエネルギーの場合には，実験結果によく合う．しかし，エネルギーが高くなると，放射の強度を低く見積もってしまい，一致しなくなる．式 (6.38) より，比例定数が ω^2 となっているので，熱い物体からは X 線など健康に有害な領域の波長が放射されているはずである．しかし，実際に観察してみるとそんなことはなく，非常に熱い物体からでも X 線や紫外線などの放射は観察されない．つまり，この式は一見正しいけれど致命的な間違いがあるということである．どこに間違いがあるのだろうか．沢山の研究者はこの問題について研究を行ったが，その答えは見つからなかった．この問題は**紫外発散**と呼ばれる．

Fig. 6.5 Examples of *standing waves* of a string at different frequencies.

Max Planck also studied the problem and was getting mad at it. Inspired by Rutherford-Bohr model of the atom, where electrons stay in a fixed orbit regulated by *quantum* numbers, one day he made the very fancy hypothesis that the kinetic energy cannot be a continuous value, but must be quantized in multiple steps of $\hbar\omega$. In this way he calculated the average energy that is not any more kT as in the equation (6.38), but a different value that drops down very fast at higher frequencies. This supposition that Energy cannot be continuous, was defined by Planck and his colleagues as an *act of desperation* ![35] How it was done ?

Planck thought that for some reason, the oscillation of electrons must be *quantized*, exactly as the strings of a guitar are forced to vibrate only to multiple frequencies. This idea was not unfamiliar to physicists, in fact standing waves in musical instruments behave exactly this way: frequency is *quantized* in multiples of a base tone (see fig. 6.5)

Planck thought then that the vibration of the molecules in a hot body, must behave exactly this way. So he simply calculated the average energy of such systems. The average energy is: $<E> = \frac{E_{tot}}{N_{tot}}$ where E_{tot} is the total energy of the system, and N_{tot} is the total number of available *states*.

---

35) act of desperation: プランクは紫外発散の問題を解決するにあたりエネルギーが連続ではないという仮説を立てたが，実はこれを信じていたわけではなかった．後に彼はこの思いつきを「act of desperation」，直訳すると「自暴自棄の行為」であったと述べている．つまり，いろいろ考えてもわからなかったのでヤケになったのである．しかし，結果的にはその説が正しかった．

6.5 黒体放射

図 6.5 弦に様々な周波数の定在波が発生しているときの例

マックス・プランクもこの問題について取り組んだ．電子が決まった軌道を持つボーアの原子模型から発想を得て，ある日彼は運動エネルギーも連続した値をとらずに，$\hbar\omega$ ごとに量子化されているのではという独創的な仮説を思いついた．この考え方を用いたところ，周波数が高くなった時の放射強度の減少がはやくなり，式 (6.38) で得られるエネルギー kT とは異なった結果を得た．このエネルギーが連続でないという仮定はプランクらによって提案されたものである．どうしてこのような発想がでたのだろうか．

プランクは，ちょうどギターの弦が振動する時，振動が振動数の倍数で起こるのと同じように，電子の振動も量子化されていると考えた．この発想は，物理学者にとって，見慣れない考えではなかった．実際，楽器に発生する定在波はこのように振る舞うからである．例えば，図 6.5 に示すように，振動数は基本の振動の倍数となり，量子化されているからである．

プランクは，熱い物体の分子の振動が，まさにこのように起こると考え，系の平均エネルギーについて計算した．平均エネルギーは $<E>=\frac{E_{tot}}{N_{tot}}$ であり，ここで E_{tot} はこの系の全エネルギーのことである．N_{tot} はとれる準位の総数である．

6.5 Black body radiation

Let's calculate first N_{tot}: lets choose one frequency ω. For this single ω, we have many possible levels of energies E_0, E_1, E_2 etcetera. Each of these has energy multiple of the minimum E_0, in this succession: $E_1 = \hbar\omega$, $E_2 = 2\hbar\omega$, $E_3 = 3\hbar\omega$ etcetera.

We will have N_0, N_1, N_2 molecules on each of these energy levels, right? The problem now is: "what will be the *distribution* of these energies?", in other words, how many molecules will be in E_0, E_1, E_2 and so on? Of course, Planck wanted to be simple and he thought at the well known Boltzmann distribution that we know very well. So he said, the number of molecules in each energy states E_0, E_1, $E_2 \ldots$, will be distributed as a Boltzmann curve, so like this:

$$N_0$$
$$N_1 = N_0 e^{-E_1/kT} = N_0 e^{-\hbar\omega/kT}$$
$$N_2 = N_0 e^{-E_2/kT} = N_0 e^{-2\hbar\omega/kT}$$
$$N_3 = N_0 e^{-E_3/kT} = N_0 e^{-3\hbar\omega/kT} \tag{6.39}$$
$$\ldots$$

where $\hbar$ is a constant. If we have an infinite number of these states, what is the total N_{tot} that we are looking for? Well, let's simplify the equation by calling $x = e^{-\hbar\omega/kT}$, then our eq. (6.39) becomes:

$$N_0$$
$$N_1 = N_0 x$$
$$N_2 = N_0 x^2$$
$$N_3 = N_0 x^3 \tag{6.40}$$
$$\ldots$$

and so on. Then the total N_{tot} is

6.5 黒体放射

まずある振動数 ω の N_{tot} について考えてみよう。この 1 つの ω に対して，どれだけエネルギー準位 $E_0, E_1, E_2 \cdots$ があるだろうか。一番小さいエネルギーは E_0 である。また，$E_1 = \hbar\omega$, $E_2 = 2\hbar\omega$, $E_3 = 3\hbar\omega \cdots$ となっている。

それぞれのエネルギー準位に $N_0, N_1, N_2 \cdots$ 個の分子があることにしよう。さて，問題はこのエネルギーの分布はどうなっているかということである。言い換えると，E_0, E_1, E_2 での分子の数はどうなっているかということである。プランクは単純でわかりやすいことを望み，よく知られているボルツマン分布を採用し，それぞれのエネルギー $E_0, E_1, E_2 \ldots$ における分子の数は，以下のようにボルツマン分布に従うと考えた．

$$\begin{aligned} &N_0 \\ &N_1 = N_0 e^{-\hbar\omega/kT} \\ &N_2 = N_0 e^{-2\hbar\omega/kT} \\ &N_3 = N_0 e^{-3\hbar\omega/kT} \\ &\cdots \end{aligned} \tag{6.39}$$

ここで $\hbar$ は定数である。もしも無限の数の準位があるとすると，N_{tot} はどうやって求めたらよいのだろう。$x = e^{-\hbar\omega/kT}$ の式を簡単にすると，式 (6.39) は

$$\begin{aligned} &N_0 \\ &N_1 = N_0 x \\ &N_2 = N_0 x^2 \\ &N_3 = N_0 x^3 \\ &\cdots \end{aligned} \tag{6.40}$$

となり，N_{tot} は，

$$N_{tot} = N_0(1 + x + x^2 + x^3 + \ldots)$$

$$N_{tot} = N_0(1 + x + x^2 + x^3 + \ldots)$$

For the theory of series it is very easy to demonstrate (simply multiply the series by x and compare it with itself) that

$$N_{tot} = N_0 \frac{1}{1-x} \qquad (6.41)$$

At this point we only need to calculate the total energy of the system. Planck simply added up the energies for each level. The total Energy at the lowest *ground* level was for simplicity set to zero, the total energy at the fist level was $N_1 \times E_1$, for the second $N_2 \times E_2$ and so on... Planck assumed that energy proceed in multiple of a base (*ground*) value $E_1 = \hbar\omega$, then $E_2 = 2\hbar\omega$, $E_3 = 3\hbar\omega$. If we proceed this way, using again $x = e^{-\hbar\omega/kT}$ the total energy is then

$$E_{tot} = \hbar\omega(x + 2x^2 + 3x^3 + \ldots) \qquad (6.42)$$

Again, if this series is infinite, the theory says that (use the fact that $x + x^2 + x^3 + \cdots \approx x/(1-x)$)

$$E_{tot} = \hbar\omega \frac{x}{(1-x)^2} \qquad (6.43)$$

so the average system energy that we are looking for is:

$$<E_k> = \frac{E_{tot}}{N_{tot}} = \hbar\omega \frac{x}{1-x} \qquad (6.44)$$

or substituting back x

$$<E_k> = \frac{\hbar\omega}{e^{\hbar\omega/kT} - 1} \qquad (6.45)$$

(note that the exponent of the exponential now is positive and not negative as usual!). This formula is what we have to substitute to $<E_k> = kT$ in eq. (6.38), it goes down very fast for higher frequency. We obtain a curve that fits perfectly experimental data:

6.5 黒体放射

となる．この式は公比 $x < 1$ の無限等比級数の和となり，

$$N_{tot} = N_0 \frac{1}{1-x} \tag{6.41}$$

となる．今の時点では，私たちは系の全体のエネルギーを計算すればよい．つまり，それぞれの準位について，単純に足し合わせればよいのである．一番低いエネルギーをゼロと設定し，$E_1 = \hbar\omega$, $E_2 = 2\hbar\omega$, $E_3 = 3\hbar\omega$ とすると，それぞれの準位での全エネルギーは $N_1 \times E_1$, $N_2 \times E_2 \cdots$ となる．プランクはエネルギーは基底状態の値 $E_1 = \hbar\omega$ の倍数であると仮定したので，$E_2 = 2\hbar\omega$, $E_3 = 3\hbar\omega$ となる．これを続け，もう一度 $x = e^{-\hbar\omega/kT}$ を適応すると，全エネルギーは

$$E_{tot} = N_0 \hbar\omega (x + 2x^2 + 3x^3 + \ldots) \tag{6.42}$$

となる．もしもこの式が無限級数だとすると，この結果は $x + x^2 + x^3 + \cdots \approx x/(1-x)$ という関係を用いて

$$E_{tot} = N_0 \hbar\omega \frac{x}{(1-x)^2} \tag{6.43}$$

となり，この系の平均のエネルギーは

$$<E_k> = \frac{E_{tot}}{N_{tot}} = \hbar\omega \frac{x}{1-x} \tag{6.44}$$

または

$$<E_k> = \frac{\hbar\omega}{e^{\hbar\omega/kT} - 1} \tag{6.45}$$

となる．ここで指数が正であることに注意すること．この式を式 (6.38) の kT の項に代入すると，高い周波数における強度の低下が大きくなり，得られた式

$$I(\omega) \propto \frac{\omega^2 <E_k>}{\pi^2 c^2} = \frac{\hbar\omega^3}{\pi^2 c^2 (e^{\hbar\omega/kT} - 1)} \tag{6.46}$$

は実験結果に非常によく一致する．

この式は，最初 ω と共に増加し，高い周波数では指数関数のほうが効いてゼロにまで落ちるということをよく表している．紫外発散は解決した．これはとても有名な，はじめての量子的なエネルギーについて考えた式であり，プラ

6.5 Black body radiation

$$I(\omega) \propto \frac{\omega^2 <E_k>}{\pi^2 c^2} = \frac{\hbar \omega^3}{\pi^2 c^2 (e^{\hbar \omega / kT} - 1)} \qquad (6.46)$$

This equation grows initially with ω as before, but at higher frequencies the exponential wins and everything goes to zero as it should be, the *UV catastrophe* was finally solved! This was the first *quantum* equation ever, it became very famous as the *Planck radiation law*, or Planck's black-body equation. The problem of the ultraviolet emission of hot body of equation (6.38) was solved for ever and *quantum mechanics* was born.

ンクの放射の法則と呼ばれる．式 (6.38) での熱い物体からの紫外線放射の問題は解決し，量子力学が誕生したのである．

Index

adiabatic（断熱），22
adiabatic expansion（断熱膨張），26
air column（気柱），56
arbitrary unit（任意単位），86

black body radiation（黒体放射），156, 158
Boltzmann distribution（ボルツマン分布），86
Boltzmann law（ボルツマンの法則），66, 70
bombardment（衝撃），98
Boyle（ボイル），4
Boyle law（ボイルの法則），6
Brownian drift（ブラウン・ドリフト），102
Brownian motion（ブラウン運動），98

Carnot（カルノー），44
Charles law（シャルルの法則），6
condensation（凝縮），124
cross section（断面積），106, 136

differential（微分），30
differentiation（微分），30, 60
diffusion（拡散），130, 148
diffusion coefficient（拡散係数），152
disorder（無秩序），44, 46
distance distribution（分子の距離の分布），72
distribution（分布），56, 88
distribution of speed（速度分布），96

distribution probability（分布確率），72
drift（ドリフト），138, 148
drift velocity（ドリフト速度），138

elastic（弾性），130
electric current（電流），144
electric resistance（電気抵抗），142
entropy（エントロピー），48, 54
equilibrium（平衡），92, 118
evaporation（蒸発），82, 116, 124

first law of thermodynamics（熱力学第1法則），44

gas（気体），90
gas phase（気相），120
Gay-Lussac law（ゲイ=リュサックの法則），6
gradient of concentration（濃度の勾配），152

heat（熱），44
heat engine（熱機関），42
Hooke's law（フックの法則），76, 86

ideal gas（理想気体），10, 92, 154
ideal gas law（理想気体の状態方程式），16
impulse（力積），12
infinitesimal（無限小），32, 34, 58
integration（積分），30
intermolecular distance（分子間距離），

88
internal energy（内部エネルギー），22, 44
isothermal（等温），20
isothermal compression（等温圧縮），20
isothermal expansion（等温膨張），20
isothermal process（等温過程），20

kinetic energy（運動エネルギー），14
kinetic theory（気体分子運動論），116

liquid（液体），90
liquid phase（液相），120

Maxwell distribution（マックスウェルの速度分布），96
mean free path（平均自由行程），104, 134, 136
mean free time（平均自由時間），104
mechanical work（仕事），44
mobility（移動度），140
molecular current（分子電流），148
molecular density（分子密度），58

Ohm（オーム），148
Ohm law（オームの法則），146
order（秩序），44, 46

particle density（分子密度），136
physical model（物理的なモデル），36
physical state（物理的な相），86
potential energy（ポテンシャルエネルギー），70, 72, 74

quality factor（Q 値），158
quantized（量子化），160
quantum（量子化），160
quantum mechanics（量子力学），166

rate of collision（衝突頻度），102, 106
Rayleigh's law（レイリーの法則），158
relaxation time（緩和時間），134
resistance（抵抗値），146
resistivity（抵抗率），146
reversible engine（可逆機関），50

second principle of thermodynamics（熱力学第 2 法則），44, 46
solid（固体），90
speed distribution（速度分布），92
standing wave（定在波），160
state（準位），160

temperature（温度），10, 14
test charge（試験電荷），142
thermal noise（熱雑音），110, 114
thermodynamics（熱力学），2, 42

universal gas constant（気体定数），16
UV catastrophe（紫外発散），158

vapour（蒸気），118
velocity of drift（ドリフト速度），138

white noise（白色雑音），48
work（仕事），50
work function（仕事関数），118, 120

索　引

【英字】
Q 値（quality factor），159

【あ】
移動度（mobility），139
運動エネルギー（kinetic energy），15
液相（liquid phase），89, 121
エントロピー（entropy），49, 55
オームの法則（Ohm law），147
温度（temperature），11

【か】
可逆機関（reversible engine），51
拡散（diffusion），131, 149
拡散係数（diffusion coefficient），153
カルノー（Carnot），45
緩和時間（relaxation time），133
気相（gas phase），89, 121
気体定数（universal gas constant），17
気体分子運動論（kinetic theory），117
気柱（air column），57
凝縮（condensation），125
ゲイ=リュサックの法則（Gay-Lussac law），7
黒体放射（black body radiation），157, 159
固体（solid），89

【さ】
紫外発散（UV catastrophe），159
試験電荷（test charge），143
仕事（work），43, 51

仕事関数（work function），119, 121
シャルルの法則（Charles law），7
準位（state），161
蒸気（vapour），119
衝撃（bombardment），99
衝突頻度（rate of collision），103, 107
蒸発（evaporation），83, 117, 125
相（state），87
速度分布（speed distribution），93, 97

【た】
弾性（elastic），131
断熱圧縮（adiabatic compression），23
断熱膨張（adiabatic expansion），23, 27
断面積（cross section），107, 137
秩序（order），45
抵抗値（resistance），147
抵抗率（resistivity），147
定在波（standing wave），161
電気抵抗（electric resistance），143
電流（electric current），145
等温圧縮（isothermal compression），21
等温過程（isothermal process），21
等温膨張（isothermal expansion），21
ドリフト（drift），149
ドリフト速度（drift velocity），139

【な】
内部エネルギー（internal energy），23, 45
任意単位（arbitrary unit），87
熱（heat），45

熱雑音 (thermal noise), 111, 113, 115
熱力学 (thermodynamics), 3, 43
熱力学第1法則 (first law of thermodynamics), 43, 45
熱力学第2法則 (second principle of thermodynamics), 45
濃度の勾配 (gradient of concentration), 153

【は】

白色雑音 (white noise), 49
微分 (differential), 31
フックの法則 (Hooke's law), 77, 87
ブラウン・ドリフト (Brownian drift), 103
ブラウン運動 (Brownian motion), 99
分子間距離 (intermolecular distance), 87
分子電流 (molecular current), 149
分子密度 (molecular density), 59, 137
分布 (distribution), 57, 89
平均自由行程 (mean free path), 105, 135, 137
平均自由時間 (mean free time), 105

平衡 (equilibrium), 93
平衡状態 (equilibrium), 119
ボイル (Boyle), 5
ボイルの法則 (Boyle law), 7
ポテンシャルエネルギー (potential energy), 71, 75
ボルツマンの法則 (Boltzmann law), 67, 71
ボルツマン分布 (Boltzmann distribution), 87

【ま】

マックスウェルの速度分布 (Maxwell distribution), 97
無秩序 (disorder), 45

【ら】

力積 (impulse), 13
理想気体 (ideal gas), 11, 93, 155
理想気体の状態方程式 (ideal gas law), 17
量子化 (quantum, quantized), 161
量子力学 (quantum mechanics), 167
レイリーの法則 (Rayleigh's law), 159

【著者紹介】

Ruggero Micheletto, Ph.D. in Physics
Born in Piedmont in Italy in 1962, he got the degree in physics in 1987 from the University of Torino and the PhD in Physics in 1992 from the University of Bologna, the oldest known university in the world. He moved to Japan in 1994. Currently he is professor of physical sciences at Yokohama City University. He does research in nano-optics, optical devices and related biological applications, experiment of visual perceptions and neural networks.

Aki Tosaka, Ph.D. in Physics
Obtained Ph.D. in physics in 2004 from Gakushuin University, in Tokyo, Japan. She then was appointed as special researcher in AIST (National Institute of Advanced Industrial Science and Technology) in Tokyo and then Research Assistant at IMRAM (Institute for Multidisciplinary Research for Advanced Materials) Institute in Tohoku University. She currently holds a full-academic position of Assistant Professor at Yokohama City University and does research on surface sciences.

* * *

Ruggero Micheletto（ルジェロ ミケレット）
1962年 イタリア ピエモンテ州トリノ生まれ．1987年 トリノ大学で物理学の修士号を取得後，1992年 ヨーロッパ最古の総合大学のボローニャ大学でPh.D.（物理学博士号）を取得．1994年に来日．現在は横浜市立大学学術院国際総合科学群自然科学系列 教授．
研究活動はナノ光学，近接場光学，光デバイスおよび知覚実験，ニューラルネットなど．

戸坂 亜希（とさか あき）
2004年 学習院大学大学院自然科学研究科物理学専攻博士後期課程修了，博士（理学）取得．独立行政法人 産業技術総合研究所 特別研究員，国立大学法人 東北大学多元物質科学研究所 研究支援者などを経て，現在は横浜市立大学学術院国際総合科学群自然科学系列 助教．
専門は表面物理学．

英語と日本語で学ぶ熱力学 *Thermodynamics in English and Japanese* 2015 年 10 月 25 日　初版 1 刷発行 2022 年 7 月 5 日　初版 2 刷発行	著　者　Ruggero Micheletto © 2015 　　　　戸坂亜希 発行者　南條光章 発行所　**共立出版株式会社** 〒112-0006 東京都文京区小日向 4-6-19 電話番号　03-3947-2511（代表） 振替口座　00110-2-57035 共立出版ホームページ www.kyoritsu-pub.co.jp 印　刷 製　本　大日本法令印刷
検印廃止 NDC 426.5 ISBN 978-4-320-03595-9	一般社団法人 自然科学書協会 会員 Printed in Japan

JCOPY ＜出版者著作権管理機構委託出版物＞
本書の無断複製は著作権法上での例外を除き禁じられています．複製される場合は，そのつど事前に，出版者著作権管理機構（ＴＥＬ：03-5244-5088，ＦＡＸ：03-5244-5089，e-mail：info@jcopy.or.jp）の許諾を得てください．